BUILDING
D·E·S·I·G·N

Improving Commercial Spaces

BUILDING
D·E·S·I·G·N

Improving Commercial Spaces

Elyse Umlauf • Phil Schreiner

**BUILDING DESIGN
& CONSTRUCTION**

The Library of Applied Design

Distributor to the book trade in the United States and Canada:

Rizzoli International Publications Inc.
300 Park Avenue South
New York, NY 10010

Distributor to the art trade in the United States:

Letraset USA
40 Eisenhower Drive
Paramus, NJ 07652

Distributor to the art trade in Canada:

Letraset Canada Limited
555 Alden Road
Markham, Ontario L3R 3L5, Canada

Distributed throughout the rest of the world:

Hearst Books International
105 Madison Avenue
New York, NY 10016

Library of Congress Cataloging-in-Publication Data
Umlauf, Elyse.
    Building design / by Elyse Umlauf & Philip Schreiner.
        p.   cm.
    ISBN 0-86636-127-8
    1. Buildings — United States — Remodeling for other
use.   2. Buildings — United States — Conservation and
restoration.   I. Schreiner, Philip.   II. Title.
NA2793.U44   1990   90-7572
720'.28'60973 — dc20   CIP

*CAVEAT — Information in this text is believed accurate, and
will pose no problem for the student or casual reader.
However, the authors were often constrained by information
contained in signed release forms, information that could
have been in error or not included at all. Any
misinformation (or lack of information) is the result of failure
in these attestations. The authors have done whatever is
possible to insure accuracy.*

Interior Design and Mechanicals by Kim McCormick

For information about our audio products, write us at:
Newbridge Book Clubs, 3000 Cindel Drive, Delran, NJ 08370

Color separation, printing and binding by
Toppan Printing Co. (H.K.) Ltd. Hong Kong

Typography by
Jeanne Weinberg Typesetting

10 9 8 7 6 5 4 3 2 1

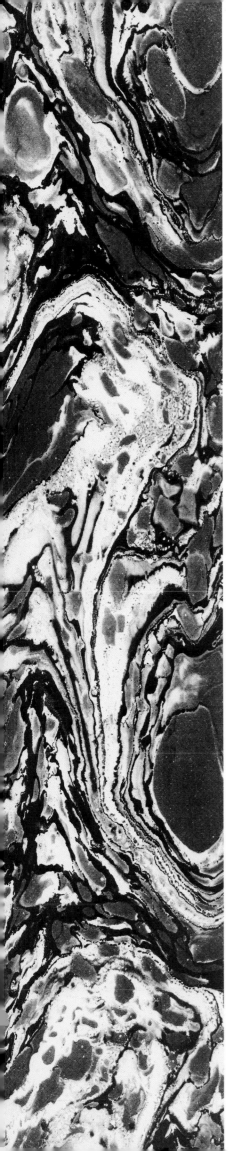

CONTENTS

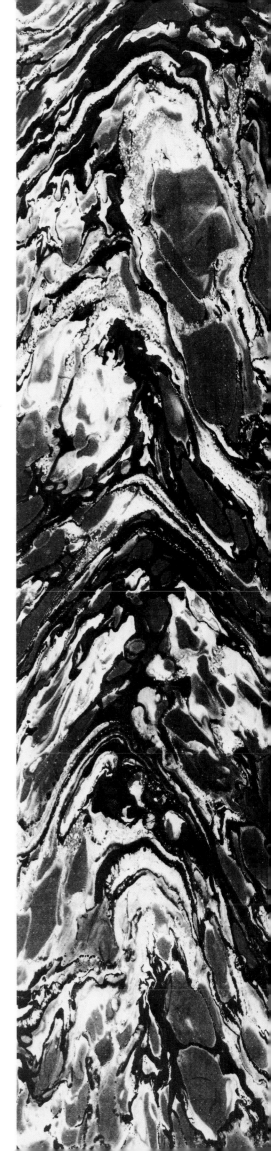

INTRODUCTION

It wasn't too long ago when renovation was considered "the poor man's market." As recently as ten years ago, renovation was the court of last resort, where development and construction teams retrenched during down times in the new construction cycles.

In those days, the availability of financing was one of the key motivators in the nonresidential, new construction market. In addition, the tax law structure was such that new construction was attractive regardless of market demand.

However, within the last decade, the construction industry has experienced many shocking, and in some cases, painful lessons. Financing is no longer the primary motivator and, regardless of market conditions, is sometimes difficult to obtain. Furthermore, more sophisticated building teams have leaned to take a serious look at other factors that combine to dictate the feasibility of new construction endeavors. These factors are far reaching, and include such sticky issues as environmental impact.

All these conditions have drastically altered the thinking of the construction industry so that, as a result, the new construction cycle has been virtually revolutionized. Building teams now look at renovation as a viable alternative, practically independent of the conditions and factors influencing trends in the new construction market.

And so, renovation has come into its own. Now the darling of developers throughout the country, the popularity growth of this market over the last decade has made renovation the new "glamour" frontier.

Renovations of older office buildings are being completed at a furious pace. As newer buildings are constructed, owners of existing structures are taking a critical look at their properties and weighing their alternatives to remain competitive. In most cases, their only alternative is renovation. If the building is basically sound, has a good location with high land value and all the other economic considerations point in the right direction, renovation is often the most sound investment. In some cases, renovation is aimed at drawing attention back to a building for marketing and/or resale purposes.

While a building's desirability can be affected by changes as subtle as a fresh coat of paint and new carpeting, a more significant attribute that can shift a structure's desirability quotient is the lobby. An atrium, landscaping and the overall quality of arrival to a building can make a big difference.

Although renovation may be a sure-fire method to draw new tenants, maintain existing ones and bring in higher rents, the overall challenges are often underestimated. Many renovations go awry and must be rethought because of ill-conceived initial plans and too little homework on the part of the renovation team.

Although we cannot save every older building (and indeed, there are many for which any type of salvation effort would be impractical), there is something magical, almost reverential, about saving worthy candidates. Successful renovation teams have learned to look at the real, functional issues and evaluate them accordingly.

That's what this book is all about: successful renovations in all segments of the nonresidential market. Renovations which were successful on a multitude of levels and which have passed the test to bring new recognition to their owners and the renovation teams who instituted the plans and carried them through to a glorious end.

In many cases, the final results are more than breathtaking, more than a monument to savvy teams who persevered, and more than a testament to creativity and patience. These buildings are models of all that can be accomplished with vision, foresight, and yes, often a lot of money. But, the preservation of classical structures, even in the case of adaptive re-use applications, is imperative if our culture is to leave its mark and message on history.

We hope you'll agree that the projects featured in this special presentation are indicative of the most exciting and creative efforts of those dedicated to the preservation of our architectural heritage.

CHAPTER

Office Building Renovation

Economics and esthetics have combined to make office building renovation a popular and lucrative alternative to new construction.

In major cities of the northeast and midwest "rust belt," the availability of vintage buildings in desirable locations has spurred a surge of reconstruction activity which has, of late, outpaced new construction activity.

In cities such as New York, Philadelphia, Washington, Boston and Pittsburgh, older office buildings are enjoying a renaissance of popularity as developers and owners rush to meet the renovation challenge. The results are often dazzling, as these "diamonds in the rough" are re-cut and polished into architectural jewels.

The charm and inherent architectural integrity of these older structures, for years ignored as the development industry carried on an exhaustive love affair with shiny new steel and glass combinations, have now come into full bloom as tastes and economics are combined to make renovation both fashionable and feasible.

Office building renovation can be either the easiest or most difficult of renovation challenges, depending on the individual circumstances of each project. No two are the same. As demonstrated by the sampling in this chapter, the movers and shakers in the construction industry take particular delight and pride in their accomplishments in the office building renovation market.

Building occupants, the ultimate benefactors of reconstruction efforts, are the real winners. They get space with charm and character, and usually at a price that makes renovated space the hottest ticket in town.

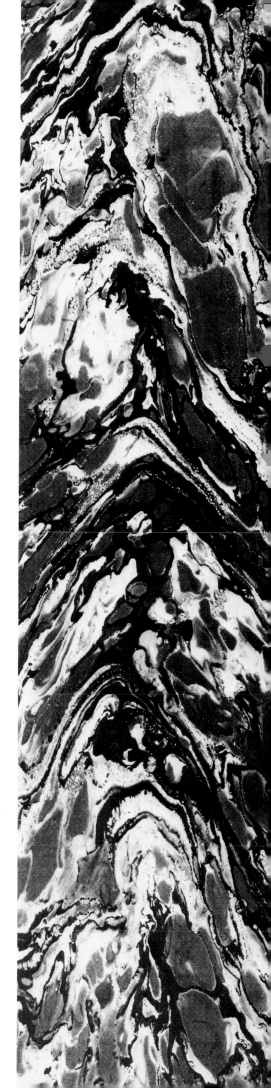

location
St. Petersburg, Florida

architect
Daniel P. Coffey & Associates, Ltd.

general contractor
Irwin Contracting, Inc.

photographer
Barry Rustin

Despite a major fire that virtually devastated a dilapidated hotel, architect Dan Coffey was able to resurrect a modern office building from the ashes and also incorporate some of the remnants left by the blaze.

The structure was so severely damaged by fire, only the exterior walls and a portion of the steel frame remained standing. While demolishing the wreck and building anew would appear to have been a good solution, it was an historic structure and the owners wanted to get the benefits of historic tax credits. The guidelines for the credit dictated that 75 percent of the original exterior walls had to be retained; the design team literally found themselves counting bricks on the charred walls to comply with the 75 percent figure.

Essentially, the reconstruction involved reframing the remains and constructing a 32,666-sq.-ft. multi-tenant office building around them. Prefabricated clearspan trusses were used to change the roof line and create a columnless third floor which optimized the available square footage. New core elements such as an elevator and exit stairs, were placed outside the old walls so as not to compromise the efficient layout of the office floors.

The new roof is covered in a lightweight aluminum tile and complements the traditional appearance of the stucco walls.

Dual entrances, one serving the street side and another serving the rear parking area, were created and linked by a central lobby. The interior is finished in light oak and the elevator lobby is made to appear lighter and brighter through the use of illuminated mirrors and oak screens.

After a devastating fire, all that remained were charred walls and a pile of ashes.

The remaining walls were reframed and incorporated into the design of the new multi-tenant office building. The tile roof was selected to complement the stucco exterior walls.

The lobby, finished in light oak, receives visitors from two sides.

TWO MELLON BANK CENTER
(UNION TRUST)

location
Pittsburgh, PA

architect
Burt Hill Kosar Rittelmann Associates (BHKR)

contractor
**Massaro Corporation-Gilbane Company
(a joint venture)**

photography
Hedrich Blessing

Of Two Mellon Bank Center, formerly known as the Union Trust Building, architectural critic Walter Kidney wrote, "It was built in a time when a first-class building was decorated as a matter of course, when an entrance lobby was a temple, a mailbox a shrine in bronze and the top of a building finished off with a flaring cornice... Fantasy is not common in Pittsburgh corporate architecture...Lucky for us to have a little of it here and there."

Unfortunately, reality had crept up on the fantasy. The building had suffered a poor exterior maintenance program, ill-conceived space reconfigurations and a tired mechanical system.

Above the four-story terra cotta mansard roof, decorated with delicate tracery and dormers, stand twin penthouse towers. Two Mellon Bank Center, formerly known as the Union Trust Building, is a well-known and well-loved Pittsburgh landmark.

Encompassing an entire city block, the Flemish Gothic building distinguishes itself from its neighbors with twin towers, a lavishly decorated terra cotta facade, stained glass, mosaic ceilings and bronze doors.

In its renovation, BHKR married beauty with utility and created Class A office space within the envelope of the 1917 landmark. The building's major tenant, Mellon Bank, occupies 700,000 square feet as well as a newly-created conference and training center at the top of the building on the tenth and eleventh floors.

The training center occupies light wells that were situated in each quadrant of the building. Originally designed to provide fresh air, light and ventilation, they had obviously outlived their usefulness. They were floored over and now house meeting space, lounge areas, break rooms and 21 classrooms.

Throughout the interior, mosaic ceilings and decorative moldings and coffers were repaired or replaced. Original colors and stenciling were meticulously restored as were marble floors, polished brass elevator cabs and bronze doors. The stained glass that caps the rotunda was cleaned and a new lighting scheme highlights its brilliant colors.

The program for the exterior involved restoring the lavishly decorated terra cotta, copying and replacing broken pieces and adding energy-efficient, but historically accurate aluminum windows.

More than 4,000 pieces of cast concrete covered with a silicone coating were used to replace damaged and missing terra cotta sections on the exterior.

Throughout the building the mosaic ceilings and all the marble and decorative moldings and coffers were cleaned and restored.

Visitors are treated to a dazzling sight upon entering the building's rotunda; the space is capped by a brilliant explosion of stained glass.

Massive bronze gates mark the
entrance to the restored
building.

Light wells that were closed over in all four quadrants of the building provide a dramatic and abundantly-lit space for Mellon Bank's training center.

THE OLD EXECUTIVE OFFICE BUILDING
SECRETARY OF THE NAVY OFFICE

location
Washington, D.C.

architect
Einhorn Yaffee Prescott

photographer
F. Harlan Hambright & Associates, Inc.

The Old Executive Office Building was constructed over a 17-year period between 1871 and 1888. At various times it housed the State Department, the Navy Department and the War Department until it became an annex to the White House in 1949. By 1930, many of the elaborate architectural decorations—wall and ceiling painting, gilded wood and marble mantles—had been obscured under more than a dozen coats of white paint.

Tests on two walls revealed that stenciled decorations and colors still existed under the layers of paint. The surfaces were stripped to the original material and the details were restored. These two walls served as the standard for color, texture and sheen coverage for the remaining wall and ceiling surfaces.

The Old Executive Office Building was constructed over a 17-year-period between 1871 and 1888. It housed various government departments until it became an annex to the White House in 1949.

The restored room is now vice president Dan Quayle's conference room.

Given the need for advanced security and communications systems, the ability for live broadcasting and the desire to accommodate mechanical replacements, original utility shafts in the walls were enlarged to provide for current and future uses. Piping was installed and capped below the floor and above the ceiling so future connections could be made without disturbing the restored space.

Prior to its restoration, much of the character of the Secretary of the Navy Office had been obscured under more than a dozen layers of white paint.

The rooms's mosaic floor is composed of mahogany, black cherry and maple woods.

The renovation called for restoring the office's stenciled walls and ceilings to their original 19th Century appearance.

575 FIFTH AVENUE

location
New York City

architect
Total Concepts/New York

construction manager
Tishman Construction

photography
Wolfgang Hoyt, Esto

Originally a furniture store and home to a flagship operation of a now-defunct merchandiser, 575 Fifth Avenue has been made over to meet the challenges of the New York of the '90s.

Inspired by the great arcades of Europe, the new contemporary retail atrium is sleek and inviting. On the outside, red granite and shiny brass trim dress up the long neglected facade. A new office tower rises above the original roof. Inside, the 70-ft.-high core space has been transformed into an exciting retail ''experience,'' despite irregularly-placed structural columns, unusually thick floor slabs and a ceiling closed off to any and all natural light.

Symmetry was accomplished through the installation of two additional columns.

The reorientation of the space as a formal entrance and hub to a network of retail specialty shops was the challenge undertaken by Total Concepts, New York. "New Yorkers, by their nature, are not the types to go into an arcade to do their shopping," says Edith Travelstead, chairman of Total Concepts. "But after traveling around Europe and experiencing the great gallerias and arcades there, we hit upon the idea of creating a 'classical mall.'"

Two additional columns were added to those in place in order to create a symmetrical colonnade; the flow of pedestrian traffic into the building was arranged so the Fifth Avenue entrance leads directly into the heart of the retail complex; floors were paved with a red-hued Italian marble to set a luxurious tone while accommodating the inevitable variety of interior shop facades and colorations.

Columns and escalator railings were hand-rubbed with lacquering material and baked with a polyester-based sealant. The finished surfaces closely approximate the color and sheen of the marble floor tiles.

The project's most memorable element hovers just three feet below the ceiling slab: a brilliant stained-glass ceiling offering tribute to the European steel-and-glass gallerias that were inspiration. The ceiling appears evenly lit, even though perimeter lighting was required for long-term maintenance. This was accomplished with specially-designed baffles for 60, 250-watt metal halide lamps.

Vertical surfaces compliment the red-hued Italian marble floors and provide an inviting ambiance at 575 Fifth Avenue.

Escalators provide immediate, efficient and virtually silent transport for shoppers within the arcade.

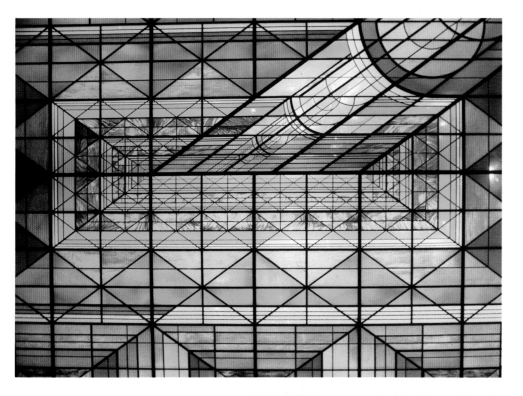

The stained glass ceiling is lighted from the perimeter through use of specially-designed baffles incorporating a Sheetrock subceiling and acrylic plates.

An artifically-lit stained glass ceiling is the perfect accent for the muted finished surfaces within the arcade. The four levels are connected via a battery of escalators plus an elevator.

A view from the top displays the dramatic panoramic view of the 70-ft.-high core space which comprises the shopping arcade.

BULOVA CORPORATE CENTER

location
Jackson Heights, Queens, New York

architect
Leonard Colchamiro Architects & Planners

owner
Blumenfeld Development Group

photography
Michael Mathers; Brian Stanton

The Bulova Corporate Center served as the world's headquarters for the Bulova Watch Company from 1953 until the early 1980s. Its close proximity to Manhattan, its 17-foot floor-to-ceiling heights and its expansive floors (the manufacturing floor alone was larger than Yankee Stadium's playing field), made it an ideal candidate for a renovation. Moreover, with its art deco design and Indiana limestone exterior, the building possessed an inherent elegance on which the architect wanted to capitalize.

Part of the reconstruction included expanding the building from 380,000 to 480,000 square feet. On the south end, a new entrance was created and its facade was produced in Indiana limestone to match the color and styling of the existing north facade.

Great care was taken during the reconstruction to preserve the existing signage, clock face and art deco facade of the Bulova Corporate Center. (Photo: Stanton)

The original north lobby contains two grand staircases finished in polished Italian marble. Its curved sections were cut from 5-inch-thick marble to yield smooth and continuous curves. Above the stairs, two 17x17-ft.-high, sculpted plaster wall murals depict the history and evolution of timekeeping devices.

A similar elegance was desired for the new 250x50-ft. common core that was cut into the center of the structure. Polished granite footpaths lead visitors along an indoor river that circulates through the length of the atrium. Terraced balconies with brushed aluminum banisters and glass walls overlook the space. A collection of plaster-cast Classical and Renaissance sculptures, on permanent loan from the Queens Museum, grace the atrium walls.

A timekeeping theme was cleverly carried into the building's restaurant, the "10:10 Club." Designed by French architect Jean-Pierre Heim, the room features a circular bar resembling the face of a watch and numerous wall clocks indicate the time in various cities around the world. A ceiling clock with large Roman numerals serves as the room's centerpiece.

The design of the new 75-ft.-high rear entrance echoes the styling and colors of the existing north facade. (Photo: Mathers)

Capped by a trussed skylight, the new three-story atrium provides comfortable gathering spots for the building's tenants. (Photo: Stanton)

Designed by French architect Jean-Pierre Heim, the time theme was incorporated into the design of the building's restaurant, the 10:10 Club. Wall clocks indicate the time in various cities around the world and the circular bar is detailed to resemble the face of a clock. (Photo: Mathers)

Among the building's amenities are a 150-seat conference room, a gymnasium and a swimming pool. (Photo: Stanton)

THE GRANT-HOFFMAN BUILDING

location
Boston, Massachusetts

architect
**Childs Bertman Tseckares & Casendino Inc.
(CBT)**

contractor
Perini Corporation

photography
Nick Wheeler

The only known example of Chicago-style fenestration in Boston is embodied in this six-story historic building in the city's Back Bay area. Clinton J. Warren designed the white terra cotta facade in 1903 and although the building was largely destroyed by fire in 1985, CBT wanted to rebuild and enlarge the building without compromising Warren's handsome facade.

Behind the facade the structure was gutted and rebuilt; a new two-story penthouse sheathed in copper was added. Its dormer windows and decorative railings were designed to complement the restored cornice and terra cotta facade.

Initially a three-story brick building, adjacent to the terra cotta facade, it was demolished and rebuilt as a seven-story structure. Molds made from original building materials allow the facade to create a visual linkage to its three-story predecessor.

The design also incorporated an adjacent three-story brick and stone building that was first demolished and then rebuilt as a seven-story structure. To preserve its integrity, the facade was disassembled, numbered, catalogued and then repaired for reuse. Exact replicas were made from the original materials to ensure that the new expansion would create a visual liaison with its three-story predecessor.

Interestingly, the two facades read as separate buildings, but on the interior, the floors extend between the two structures and they are actually one building with a single address. The interior is composed of office and retail space.

While the distinct facades may read as two separate buildings, on the interior this is actually one building.

The addition of a two-story copper penthouse captured an additional 16,000 square feet for office space above the restored terra cotta facade.

FRONT ELEVATION

739-749 BOYLSTON STREET

1500 K STREET

location
Washington, D.C.

architect
The Weihe Partnership

finish consultant for public areas
Keyes Condon Florance

contractor
OMNI Construction, Inc.

photographer
Dan Cunningham

In its heyday, the headquarters of the Southern Railway Company, now known as 1500 K Street, was one of Washington's grandest buildings. Constructed in the 1920s, its bronze storefronts on the ground-level floors, carved columns, stone cornices and patterned zinc spandrels reflected the Federalist architecture popular throughout the city during that period.

In the original lobby, fresh paint in shades of green and gold draw attention to the highly ornamented plaster ceiling. Second story balconies, accentuated by bronze and marble, contribute to the visual interest of the space. The original black granite floor was cut in a ring to surround a new patterned floor composed of four marbles—calacatta vagli, negro marquina, verda antique and cherry red. The new marbles, authentic to the period, harmoniously coexist with the original gray-beige marble walls.

The 60-year-old limestone exterior was completely cleaned and repaired. The bronze storefronts and exterior metal finishes were all refinished.

On a long blank facade fronting a busy thoroughfare, K Street, a new opening was cut to allow visitors access to the building through dual entrances. Its marble floor matches the one installed in the existing lobby and serves as a unifying element between the old and new spaces. The ''L'' shape created by the double lobbies is further linked by a two-story fountain court, detailed to appear as if it is an original feature of the structure. The marble floors from both lobbies come together in a geometric design in the court. The fountain serves as a centerpiece for the entire area.

During the renovation, an additional 27,000 square feet was captured by infilling part of an existing courtyard space from the second floor up. Extensive cleaning and repairs infused new life into the limestone exterior and double-glazed windows set in forest green frames further improved the stately facade.

The dilapidated windows were replaced with double-glazed, thermal break windows set in forest green frames.

A new fountain court adjacent to the elevator banks was detailed to appear as if it had been an original feature of the building. A marble fountain, accented with bronzed spouts in a lion head motif, is the centerpiece of the area.

The top of the building and its rooftop terrace is dramatically illuminated at night.

Craftsmen cleaned and repainted the 15th Street lobby's ornate coffered ceiling in shades of green and gold. Intricate marble carvings and cornices and bronze accents also grace the two-story lobby.

MELLON INDEPENDENCE CENTER

location
Philadelphia, Pennsylvania

architect
Burt Hill Kosar Rittelmann

photography
Tom Crane Photography, Inc.

One project manager compared the building to a Hollywood set: It had a beautifully-detailed facade, with virtually nothing behind it. Formerly the Lit Brothers Department store, the building had housed one of Philadelphia's major retailers until the store closed in 1977. The 12 interconnected buildings—constructed between 1863 and 1907—were rescued from a number of dates with the demolition contractor. In 1985, the seven-story landmark was purchased by a developer who had Mellon Bank committed as a major tenant.

The 900,000-sq.-ft. building had so steadily degenerated that huge sections of the roof had collapsed two to three stories down into the structure. However, within 24 months, the building was transformed from its ramshackle state into attractive office and retail space.

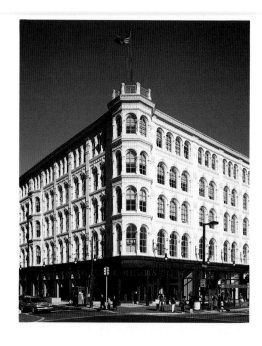

One of the more significant challenges involved leveling the uneven floors that resulted from the various infull projects and renovations that were undertaken during the building's history. As the department store grew and required more square footage, the space was expanded by purchasing adjacent structures and connecting them to the existing buildings with ramps.

Reconfiguring the interior for commercial success involved cutting a 40-ft.x60-ft. atrium at the center of the building that extends through the full seven stories. Besides creating a grand space and bringing natural light inside, the atrium also helped to break up the enormous floor plates into manageable office space. The atrium is handsomely finished with coffered ceilings, oak trim and a blue, gray and white marble floor. Mellon occupies 650,000 square feet, while 235,000 square feet in the atrium is dedicated to retail stores.

The block-long facade of the complex had also become quite an eyesore. Since the buildings had been constructed during different periods, the facades were made of a variety of materials, including cast iron, brownstone, marble, terra cotta and brick. All had deteriorated significantly and some were covered under 25 layers of peeling paint.

Beyond repairing damaged ornamental pieces, the principal challenge was visually unifying the disparate facades. This was accomplished by applying a special white paint that provided a uniform color and texture over all the assorted exterior materials. The Mellon Independence Center's facade is now listed on the National Register of Historic Places.

Offices are dispersed around
the upper stories of the atrium
and its two first levels are
enlivened by a number of
retail stores and restaurants.

The facade of Mellon Independence Center presents a visually unified appearance despite the fact that it is composed of numerous materials, including cast iron, marble and terra cotta.

After the Lit Brothers department store closed in 1977, the store was under an almost constant threat of demolition until its renovation in 1985.

COMMERCIAL NATIONAL BANK BUILDING

location
Washington, D.C.

architect
Keyes Condon Florance

contractor
OMNI Construction, Inc.

photographer
Dan Cunningham

When Waddy B. Wood's Commercial National Bank Building was constructed in 1919, its design epitomized the Washington, D.C. style—a neo-classical building emulating the capital's monuments and national buildings. When the building opened, architectural critic Leon Sloan praised it as a well-designed structure and one in which the architect was "singularly successful" and "...used economy as a spur to achieve a simple elegance."

In the ensuing years, however, the 11-story structure had been shamefully maintained and had steadily deteriorated into a gloomy hulk, presenting a harsh contrast to the graceful, historic buildings in its environs.

Two 30-ft.-high Doric columns dominate the dramatic 75-ft.-long Grand Banking Hall. Its decorative features include a marble floor and a coffered plaster ceiling.

An extensive restoration reclaimed the building's eclipsed grandeur and a 197,742 square foot addition—physically and esthetically linked to the original building—more than doubled the square footage available to new office tenants.

With minimal historical documentation to rely upon, the restoration evolved into something of a detective story. The banking lobby displayed a hung ceiling only about eight feet above the ground, yet the permanent ceiling was 30 feet high. After removing layers of lay-in ceiling and ductwork, an obscured mezzanine level that overlooked the grand banking hall was discovered as well as a significantly damaged coffered ceiling. Two 30-ft.-high plaster columns had been similarly concealed behind a purple-painted drywall.

Several of the four-square ceiling coffers, irreparably damaged were recast and replaced, while the salvageable plaster and decorative details bordering the ceiling were either restored or replaced with moldings cast from originals. By using a marbelizing technique, the two fluted Doric columns were restored to resemble antique verde. The columns dominate the far end of the bank lobby and frame the polished bronze details below the mezzanine balcony.

The ceiling coffers and wall freizes were dressed in shades of blue, gold and rust and a decorative surround was added around the original entrance. The marble floor, severely damaged, was removed and replaced. A salvageable portion of the original portoro (a black marble with gold veins) was skillfully blended with five other marbles in patterns and colors harmonious with the ceiling's design and palette.

Installations of ductwork and suspended ceiling systems had significantly damaged the coffered ceiling and much of the building's elegant finishes.

Prior to the restoration, this mezzanine balcony overlooking the lobby had been obscured by a hung ceiling.

Decorative lion head gargoyles at the eleventh floor of the existing building were removed, cleaned and remounted. The detail also appears on the exterior of the addition to provide a visual link between the two structures.

The new New York Avenue facade effectively integrates itself with the other historic buildings surrounding it.

The colors and patterns on the coffered ceiling and marble floor were selected to echo one another.

The original facade at 14th and G Street was chemically cleaned and tuckpointed. An 11-story addition on the New York Avenue side created an additional 197,742 square feet of leasable area.

FEDERAL RESERVE BANK

location
Chicago, Illinois

architect
Holabird & Root

contractor
Pepper Construction Company

photography
**Steinkamp Ballogg (exterior);
Nick Merrick (interiors)**

Although the Federal Reserve Bank's workload and services expanded considerably, its space had not. Despite the fact that the downtown bank occupied 820,000 square feet in two buildings constructed in 1921 and 1957, it was still suffering a severe space crunch. The bank's management realized that to maintain an ageless, dignified image and establish an environment comparable to its competitors, it would have to undertake an extensive building program.

The $89 million plan involved the addition of 165,000 square feet and a renovation of all the existing space.

The renovation was driven by the need to cluster similar banking functions together and reorganize the vast 50,000-sq.-ft. floor plates into visually unified, but open-plan work spaces. To avoid a confusing "rat's maze" of cubicles,

The most challenging aspect of the exterior renovation was visually reconciling the limestone facade of the original 1921 building and the metal-clad 1957 addition with the new construction.

workstations were dispersed along an east-west axis, with 48-in.-high and 68-in.-high panels dividing them. Private offices and a conference room are clustered toward the core of the building and along the north and south walls. To introduce natural light and an airy feeling, two atriums were opened on the west side of the building. The tree-filled spaces step back from floors two through seven and again from floors ten through sixteen. So visitors are clearly oriented, each elevator lobby corridor leads to an atrium on one side and a reception desk on the other.

A standardized color scheme helps to visually unify the office floors and makes interchangeability of furniture or workstation panels possible. All the floors include off-white walls, gray carpeting, wood panels and trim and black railings. Three accent colors—burgundy, blue and hunter green—are repeated throughout the building.

Equally as important as internal organization was creating a hospitable environment for visitors to the facility. Prior to the renovation there were dual entrances to the bank—a more formal public entrance and another that was frequented by employees. In the redesign, a second-floor mezzanine was removed to create a dramatic four-story lobby elegantly finished in marble and granite. Everyone now accesses the building from one main entrance. Acoustical ceiling tiles were removed and the original neoclassical decorations were restored and painted white and gold. The space is illuminated by bronze wall sconces and existing elaborate chandeliers that were cleaned and restored.

The lobby's clean straight lines, rows of marble columns and brass railings all help to create the subtle elegance that the bank's management envisioned at the outset of the project.

So visitors are clearly oriented, each elevator lobby features a reception desk on one side and the atrium balcony on the other.

Brass wall sconces in the entrance lobby draw the eye upward and highlight the lower portions of the marble columns.

A dramatic, stately lobby creates a hospitable welcome for visitors to the building.

Stacking Diagram
Federal Reserve Bank
Chicago, Illinois

This diagram illustrates how the interior space was organized to cluster similar banking functions together. Office functions are situated on the upper levels while high-security areas are located on the lower levels.

The programming process for the Federal Reserve Bank was an all-encompassing effort to organize the 2,000 employees and 50 departments of the Bank. The new organization concept places the operational, high security areas of the bank in lower levels, while more typical office functions (which are more viable) are located at the upper atrium levels. Computer and check processing operations are removed from any atrium space, while the lower atrium ends with an expanse of space at the employee cafeteria.

Function		Level
Auditors		16
Supervision & Regulation	Loans & Reserves	15
Supervision & Regulation		14
Financial Management	Support Services	13
Operations		12
Research		11
Computer Support		10
Computer Operations		9
Check Processing		8
Check Support		7
Computer Support		6
Cafeteria	Conference Center	5
Support Services		4
Customer Services / Officers Dining		3
Legal / Exec. Area		2
Lobby / Government Securities		1
Secure Operations		1B
Secure Operations		2B
Secure Operations		3B

All the office floors contain open-plan workstations with a combination of 48-inch-high and 68-inch-high partitions.

Two new atriums were created on the west side of the building between floors two through seven, and between floors 10 through 16.

CHAPTER

Adaptive
Re-use

The general public first became aware of adaptive re-use when abandoned and largely ignored warehouses, often situated in distinctly un-tony sections of our cities, burst forth as revamped, chic loft residences.

The latest steel and glass box often inspires a yawn or a sigh, while these seemingly has-been buildings can evoke a gasp. As more of them are transformed into spectacular commercial, corporate and retail spaces, the market and the public become hungrier for anything old turned new—former factories, warehouses and even churches.

The designers of these conversions possess a unique genius, vision, creativity and the ability to adapt to a constantly changing project scope. Sometimes spectacular decorative elements are unveiled and quickly incorporated into the renovation strategy. At other times, hopes for a particular design component are nixed because of existing conditions.

Adding to the challenge is an array of physical constraints such as predetermined square footage, awkwardly-placed windows, or oddly-configured space.

These architects are not overly-preoccupied with perfection, given the fact that flaws, rough edges and whimsy enhance the building's character. Yet, they also walk a fine line in trying to develop a unifying theme without creating an environment that seems contrived or trendy.

The projects in the following chapter are magnificent illustrations of buildings that were transformed into treasures from what the untrained eye would have considered trash.

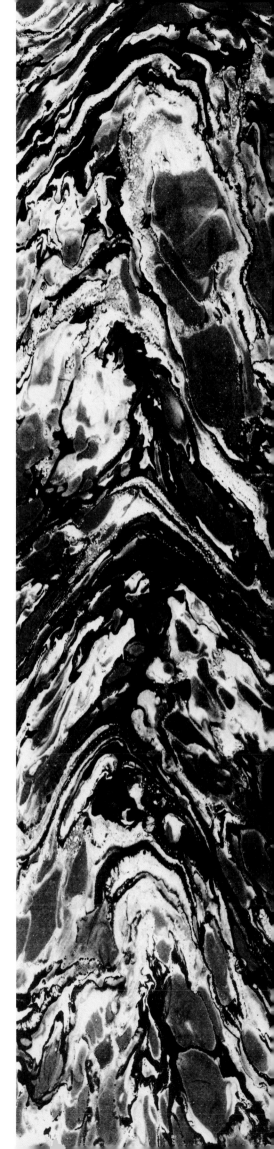

CHARLESTOWN NAVY YARD/ BUILDING 149 & BUILDING 199

location
Boston, Massachusetts

architect
**Huygens DiMella Shaffer and Associates, Inc.
(shell buildings); Jung/Brannen Associates, Inc.
(laboratory)**

construction manager
Morse/Diesel

developers
**The Congress Group, Inc. (shell buildings); Navy
Yard Biotechnical Research Associates (for the
Lawrence E. Martin Laboratory, a research
component of the Massachusetts General
Hospital)**

photography
**Abbott Boyle (before); Nick Wheeler (renovated
interiors and exteriors); Richard Mandelkorn
Photography (laboratory space)**

The Charlestown Navy Yard is a 130-acre commercial, residential and recreational development that was previously the central location for the construction and repair of U.S. Navy ships. Shipbuilding operations ceased when the compound was closed in 1974, but it still is the permanent home of "Old Ironsides" —the USS Constitution.

This project was a renovation of two Navy warehouses, one built in 1917 and the other constructed in 1945. The inherent challenge of the project involved adapting the 1.3 million square feet to be appropriate for a mixed-use facility, while maintaining the industrial character of the structures.

Renovated under the guidelines of the National Park Service and the Boston Redevelopment Authority, Building 149 now houses 700,000 square feet of office and retail space and Building 199 provides parking for 1,360 cars.

*A new multiple-mullioned
aluminum window system
was developed to replicate
the profiles and sizes of the
original steel windows that
had severely deteriorated.*

The massiveness of Building 149 dictated a design solution that would bring natural light and a human scale to the 400-ft.-long interior. The introduction of two skylit atriums solved both problems, and allowed the architect to recycle some existing building components: loading docks became pedestrian walkways and restaurant terraces, and bay windows, balconies and planters were sculpted from the massive concrete supporting columns.

On the exterior, the existing steel windows had deteriorated beyond repair; a multiple-mullioned aluminum window system was designed to replicate the profiles and sizes of the originals. On Building 199's exterior, a steel and glass elevator tower was attached and provides riders in the glazed cab with spectacular views of Boston's harbor.

The major tenant in Building 149 is the Lawrence E. Martin Laboratory, a biomedical research facility occupying 380,000 square feet. While it was a significant challenge to meet the extensive demands of a flexible laboratory within the building's tight floor-to-ceiling heights, Jung/Brannen Associates was able to insert all the necessary services by devising a creative layout that maximized the available square footage.

As it turned out, the building was ideally-suited for the laboratory's needs because the rigid frame provided a vibration-free environment that was extremely appropriate for sensitive scientific equipment. Approximately 40 percent of the facility is occupied by laboratory bench areas and the rest consists of animal facilities, meeting rooms and libraries. The space was laid out so that lab benches could be situated around the perimeter of the building to provide researchers with views out to the city and harbor. Lab support spaces were built toward the interior; offices overlook the eight-story atrium.

Designing human-scaled space within Building 149's vast interior provided the architect with a significant challenge.

Pedestrian bridges link the office building and parking garage. A glass elevator tower attached to the exterior of the parking garage offers riders in its glazed cab with spectacular views of the harbor.

Building 149's major tenant is a biomedical research facility occupying 380,000 square feet.

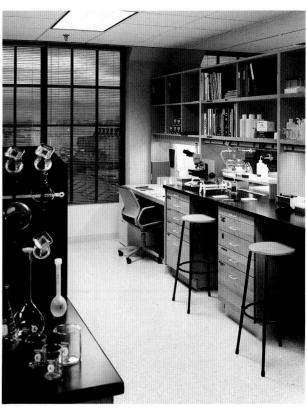

Within the 400-ft. x 100-ft. interior of the building, two skylit atriums were created. Existing building components, such as the loading docks recycled into pedestrian walkways, were incorporated into the new design.

ENGINE COMPANY NUMBER 28

location
Los Angeles, California

architect
Altoon & Porter

general contractor
Lusk Construction Company (shell building)

photographey
Glen Allison (interior); Berger Conser
Architectural Photography (restored exterior)

Engine Company Number 28, a three-story reinforced masonry building constructed in 1912, had outlived its usefulness as a fire station. The building sat vacant for nearly 16 years until it was designated an historic landmark and converted into office space.

The original facade was designed in a Renaissance Revival style, with distinguishing features such as twin parapet towers with inset arches and two-story arched bay windows.

The existing 20,000 square feet was thought to be too small to make the adaptive reuse a commercial success. Moreover, historic preservation guidelines dictated that any addition be clearly distinguishable from the original building. To gain additional space, Altoon & Porter designed a 14,000-square-foot, three-story expansion at the back of the property and added an entire new fourth floor on both the existing building and the expansion.

Constructed in 1912, the fire station was among the first reinforced concrete stations, and in its day was considered state-of-the-art.

The new facade echoes the original front facade in its proportions and shapes, but because it was executed in contemporary materials, it differentiates itself from the historic structure.

Skylights on the original top floor were removed, leaving a large opening. A new skylight, built on the roof of the fourth level, helps to create an abundantly-lit atrium between the third and fourth levels.

The truck garage, with 16-ft.-high ceilings, was cleverly reconfigured to house a restaurant. Its existing decorative metal ceiling, brick floor and walls of green Italian tiles were incorporated into the eatery's decor.

Wherever possible, the character of the fire house was retained by leaving original elements such as the fire pole and an iron stove in place. The massive fire doors were also salvaged and placed in the open position at the lobby entrance.

Although historic preservation guidelines would not allow changes to the facade, the interior, vacant for 16 years, required extensive renovation to make the building a commercial success.

The original architectural firm, Kremple & Erekes, designed the building in a Renaissance Revival style and decorated the facade with two-story arched windows and terra cotta cartouches depicting firemen's helmet and tools and the seal of the City of Los Angeles.

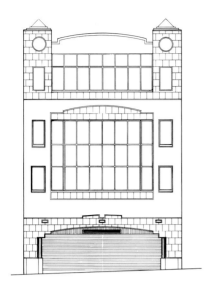

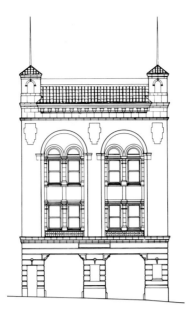

While the new addition (left) echoes the design of the original facade (right), it was executed in contemporary materials and clearly distinguishes itself from the existing building.

LEBANON ST. ELEVATION   FIGUEROA ST. ELEVATION

An abundantly-lit atrium space was created between the two top floors.

Interior materials such as wood, marble and glass help to create a contemporary interior within the envelope of the historic building.

THE LANDMARK BUILDING

location
Boston, Massachusetts

architect
Jung/Brannen Associates, Inc.

contractor
Turner Construction Company

photographer
Nick Wheeler

Formerly known as the United Shoe Machinery Building, the 447,000-square-foot Landmark Building is Boston's largest art deco structure, and its renovation represents one of the most extensive ever undertaken in the city.

With its ziggurat massing of yellow-beige brick, ornamental cast stone and gold-tiled roof, the building was the first architectural response to the 1928 Boston Zoning Law, which aimed to preserve light and human scale at the street level. The 24-story building, constructed in 1929, is stepped back from the street at an angle of approximately 68 degrees, allowing air and light to reach the ground.

Constructed in 1929, The Landmark Building is Boston's largest art deco building.

On the exterior, 1,800 pieces of cast ornamentation in 60 decorative motifs—fountains, pineapples, spirals and flowers—were recast. The truncated pyramid roof of glazed orange-yellow tiles, was cleansed of the accumulated grime that had dulled its gleam. Over 1,000 energy-efficient insulated windows were replicated to match the original steel-framed windows.

With gray-green and black marble walls and chandeliers with elaborately etched glass panels, the Landmark's lobby is one of the city's most spectacular examples of art deco design.

The cast bronze elevator doors, engraved with a tree-of-life motif, were cited by the Boston Landmarks Commission as one of the structure's most outstandng examples of art deco decoration. To accentuate their beauty, the elevators' bronze door frames were reconstructed to incorporate a continuous band of narrow light. The lights dramatically enhance the doors' beauty that previously had been lost in shadow.

The facade, illuminated in the evening, received the Illuminating Engineering Society of North America's Outstanding Lighting Design Award for facade lighting. The Landmark's other recognitions include the National Historical Preservation Award and the Massachusetts Historical Preservation Award.

A continuous band of light was added to the elevator door frames to accentuate the cast bronze doors.

The building's lobby is considered one of the most spectacular examples of art deco design in the city.

The massing of the 447,000-square-foot structure is stepped back to preserve light and a human scale at the street level.

The exterior restoration restored the gleam of the pyramid roof which had been dulled by decades of accumulated grime.

More than 1,800 pieces of cast ornamentation in 60 different decorative motifs were accurately recast to replace damaged sections.

RIVERFRONT PLAZA

location
Milwaukee, Wisconsin

architect
Herbst Eppstein Keller & Chadek, Inc.

construction manager
Hunzinger Construction Company

owner
PLB/LW Partnership

photography
Edward J. Purcell

The six-story Riverfront Plaza is a significant component in the rebirth of downtown Milwaukee's northern fringe. Located on an historic block —Old World Third Street—the building is bordered by the Milwaukee River on its east elevation, which also makes it an important link in the city's riverwalk system. The non-descript warehouse was converted into 100,000 square feet of class "A" office space and 20,000 square feet of restaurant and retail space.

It was desirable that the redesign of the exterior be sympathetic to the character of the street, but still provide the building with a fresh image. This was accomplished with a new window scheme comprised of large, half-round windows that punctuate the upperportion of the facade. The new exterior skin consists of brick, limestone and an exterior insulation system. The first floor contains a limestone base, new brick piers, insulated glass and fabric awnings.

The new mixed-use building is an integral part of the rebirth of downtown Milwaukee's northern fringe.

On the north facade of the building a porte cochere defines the entrance and provides a sheltered drop-off for visitors to the building. An existing two-story train bay was deftly redesigned into a central core and a two-story lobby for the offices above. The space is finished in mahogany, marble and brass. The ground level also includes retail space and a well-positioned restaurant that offers its patrons views of the river.

A subsequent phase of the project will include the completion of an outdoor riverwalk that will link Riverfront Plaza with the other buildings along the historic block and the emerging city-wide riverwalk system.

Prior to its renovation, the building was just another neglected, nondescript warehouse structure.

An existing two-story train shed was skillfully converted into a central lobby finished in mahogany, marble and brass.

The new exterior skin is essentially composed of limestone and brick. A new window scheme helped the building to project a contemporary image, while harmonizing with the neighboring historic structure.

Offices are provided with views of the city through the wide, half-round windows that punctuate the upper floors of the center.

The second phase of the Riverfront Center project will involve the development of a riverwalk to capitalize on the structure's choice location along the Milwaukee River.

THE GRAND BALDWIN

location
Cincinnati, Ohio

architect
KZF, Inc.

owner
Corporex Properties, Inc.

photography
**Wayne Cable (interiors); Ben McIntyre (before);
Jeff Friedman (exterior)**

The adaptive reuse of the former Baldwin Piano Factory defied the skeptics who doubted that a functional office environment could be created among the long, narrow floors of the 66-year-old building.

Given the structure's 400-foot length, it required an imaginative design to introduce well-scaled office space in what seemed to be an overwhelmingly vast environment.

KZF was one of the Grand Baldwin's earliest tenants, occupying the two top floors of the building. On the eighth floor, attention was paid to creating as non-hierarchical an environment as possible. Surrounded by sloping windows, it was conceived as a large atelier space for the design and drafting room. Manager's workstations are dispersed along the perimeter, but panels separating them are primarily clear glass to allow unobstructed views from essentially anywhere in the room.

Formerly the Baldwin Piano Factory, this 66-year-old building was skillfully converted to speculative office space.

Executive offices, the boardroom and an art gallery are housed on the seventh level. The reception area's floor is covered with parquet-patterned granite composed of tan niagara and beige inca.

A 100-ft.-long barrel-vaulted circulation spine connects the reception area to the boardroom and executive offices. Artwork placed on the walls helps to minimize the feeling of walking down an endless corridor. Its length was further broken up by placing columns along the walkway, framing the art pieces and helping to subdivide, visually and psychologically, the thoroughfare into more manageable segments. Indirect lighting provides a soft ambient effect well-suited for the art pieces. The subtle pindot pattern of the carpeting echoes the parquet pattern of the granite floor in the reception area.

KZF's space serves as a model for the building's other tenants, suggesting how a suitable environment can be created in seemingly awkward conditions, as well as a showcase to display the diversity and skill of its design staff to potential clients.

Given the long, narrow floorplates, skeptics doubted that the raw manufacturing space could be converted to practically-scaled office space.

KZF occupies the two top floors of The Grand Baldwin. Arched windows in the seventh floor reception area provide visitors with a panoramic view of Cincinnati's skyline.

The barrel-vaulted walkway connects the reception area with executive offices. Its length is broken up by well-situated columns and artwork placed on the walls.

The eighth floor houses the design and drafting space. Largely designed in an open office configuration, unobstructed views out the sloping window are available from virtually anywhere in the room.

The renovation of The Crown Roller Mill pays homage to the history of the building, while simultaneously providing 130,000 square feet of contemporary office space.

To retain the historical integrity of the structure, additions and appendages were avoided in the design of the reconstruction. A new structural frame was inserted into the shell that was left by the fire and the original walls were partially rebuilt and incorporated into the design. The mansard roof was recreated and the new windows match the scale and proportions of the originals.

Through the selection of materials, textures and colors, a relationship between the interior and exterior was established, A fountain in the center of the space was designed as a symbolic reminder of the significance that water had to the previous functions of the building. The water cascades continuously from the second floor down to the first level.

In 1983 a fire gutted The Crown Roller Mill. With hope of reconstructing the building, the city braced the brick and limestone walls that were left standing.

A fountain at the center of the space recalls the significance that water had to the building when it functioned as a flour mill.

A new plaza and extensive landscaping enhances the appeal of the building and its location along the Mississippi River.

A mansard roof, faithful to the appearance of the original one, was sheathed in copper.

HAINES LUNDBERG WAEHLER OFFICE

location
New York, New York

architect
Haines Lundberg Waehler (HLW)

general contractor
Rampart Construction Associates Inc.

photography
Peter Paige

This was only HLW's fifth move during the 103 years the architectural/engineering firm had been in practice. For their new offices, HLW's managers desired a light, airy environment. Since the firm would occupy the quarters for many years, flexibility to reconfigure was desirable. One of the key criterion driving the design was the requirement that the space would foster communication among the HLW staff.

The firm opted for two floors in the former Arnold Constable department store, built in the city's Chelsea section in 1877. The block-long fifth floor, once the sales floor for women's foundation garments, featured tall, arched windows, high ceilings and corinthian columns, while the less desirable sixth level had comparable space but small windows situated more than six feet above the floor.

Two original cast iron columns were incorporated as design elements in the circular reception area.

An interior "street" running through the entire length of both floors became the main organizing element of the space. It segregates the long floors into manageable segments and, by virtue of the layout and location of the various services, the street also tends to encourage interaction among HLW employees. Design studio space, reconfigurable partitioned workstations and private offices are located on the windowed side of the street, while support services such as conference room. the reception area and the library are located on the other side.

The essential logic behind the plan is that when employees cross the street through the course of the day, they are likely (just as they would be on a city street) to bump into co-workers and stop to chat. The tree-lined streets are equipped with benches and comfortable chairs to encourage impromptu meetings or design critiques. The sixth floor is similarly configured, although the work areas were raised on precast concrete pedestals to allow views out the windows.

With high ceilings, tall arched windows and Corinthian columns, the block-long fifth level provided an ideal environment for HLW's new design studions.

The interior "street" running the length of the both floors is the main circulation spine.

Department heads and project managers occupy private offices adjacent to the departments they oversee.

Architectural, engineering and interior design department workstations are dispersed just off the street. Existing columns throughout the space, painted white, are remnants from the existing building.

This is one of several
conference rooms used for
formal gatherings between
clients and the HLW staff.

While the majority of the space is open offices, managing partners, department heads and project managers all received enclosed private offices. Pictured here is one of the partners' quarters.

ALBANY INTERNATIONAL CORPORATE HEADQUARTERS

location
Albany, New York

architect
Einhorn Yaffee Prescott

photography
Bill Murphy

Founded nearly 100 years ago, Albany International is a designer and producer of paper machine clothing. An historic complex of buildings dating from the late 19th century had housed its northeastern United States operations and manufacturing headquarters until 1986 when a new manufacturing facility was built at a different site.

EYP was retained to update the complex to accommodate a modern, efficient headquarters.

Comprised of over 450,000 square feet, the buildings, with their ornate brickwork, were architecturally appealing and a faithful representation of the architecture of late 19th century mills.

The Albany International Corporate Headquarters consists of a series of buildings dating from 1902. Shown here is the south facade overlooking a pond on the site.

The site's first building, a brick stair tower and the three-story factory building, was constructed in 1902 and numerous additions were made between 1910 and 1948. The last office building was added in 1954.

Creating a single, functional unit out of the various buildings was one of the more challenging aspects of the project, which was accomplished by rebuilding the original core of the complex.

The structure was gutted and a monumental stairway was built within the original stair tower to provide vertical and horizontal circulation throughout the mill and the office buildings flanking the tower. The stairway also allows the office and its later additions to function as a single unit.

A new entry pavilion, situated at the base of the tower, functions as a reception area and is the focal point of the entire complex. Materials—granite, brick, copper roofing and wood windows—were selected to complement the character of the mill.

Similarly, on the interior, finish materials were sensitively integrated to complement the original design. Original wood panelling was restored in the president's office and corporate boardroom and new lighting fixtures, authentic to the period, blend with the structure's fabric. Custom mahogany millwork, veneer panel wainscot and custom-built credenzas further emphasize the historic motif.

Mahogany-framed photos dating back to 1890 were placed along the corridor walls, exhibiting the company's history and growth.

The entry pavilion at the base of the tower is clearly identified as a new design element but its architectural materials—granite, brick, wood windows and copper roofing—are sympathetic to the character of the mill.

A monumental staircase, built within the original tower, provides a circulation spine for the rebuilt complex.

The reception area serves as a
visual focal point for the
building complex as well as
an unobtrusive security hub.
The reception desk and
security controls at the main
gate monitors various points
situated around the grounds.

215/217 FIFTH AVENUE NORTH

location
Nashville, Tennessee

architect
Gobbell Hays Partners, Inc.

photography
**Rion Rizzo (Creative Sources
Photography/Atlanta)**

Within the confines of two turn-of-the-century buildings, architectural/engineering firm Gobbell Hays Partners created new offices for itself.

The two buildings, 215 and 217, had been used for a variety of purposes over the years, including a photography studio, an ice cream parlor and a ballroom dancing school.

At the street level, the buildings are divided, with the 215 side occupied by an office supply and printing store. The 217 side serves as the entrance to GHP's quarters.

In the interior, the second and third levels of the two buildings were united and GHP occupies the entire third floor. What was once a ballroom is now the firm's library and reception lobby. A fireplace, existing moldings, wood floors and exposed brick walls make the space visually enticing.

Beyond tuckpointing and weatherproofing no major modifications were made to the exterior of the two buildings.

While retaining the hard surfaces such as brick and wood was desirable from an esthetic standpoint, it created a noisy workplace. The problem was alleviated by installing a fiberglass ceiling that absorbs and stops sound from bouncing throughout the space. Oriental rugs scattered around the office serve the dual purpose of absorbing noise and lending an air of coziness.

Adapting the buildings to new uses necessitated replacing the steel-framed skylight over an open-area conference space with a more energy efficient aluminum one. A new skylight was also added in the main conference room. The theatrical lighting throughout the workspace is bounced off the ceiling to create the indirect effect necessary for comfortable computer use.

The main lobby off the street is marked by leaded glass doors. The moldings in the lobby and throughout the building are poplar, with a mahogany stain.

What was once a ballroom is now Gobbell-Hayes' library and reception area.

This stairway is the circulation link between the two buildings. The oak banisters are an original feature of the building.

Theatrical lights, reflected off the ceiling and walls provide indirect lighting ideal for computer uses.

Arched windows, exposed
brick walls and oak flooring
make the space extraordinarily
appealing.

1010 METRODOME SQUARE

location
Minneapolis, Minnesota

architect
Setter, Leach & Lindstrom Inc.

contractor
Kraus-Anderson Construction Company

photographers
George R. Heinrich Photography (interiors)
Phillip Prowse Photography (exteriors)

Previously, this 60-year-old building was home to the Strutwear Knit Company. An extensive reconstruction greatly elevated the status and infused new life into the previously unprepossessing factory building situated on the fringe of Minneapolis' downtown.

Within the U-shaped configuration of the original structure, the architect successfully infilled a five-story, skylit atrium between the protruding arms, creating a new entrance in what was originally the back of the factory.

The new entrance pavilion and atrium serve as a public "street" and gathering place within the building. Circulation bridges connect the wings of the "U" at every level.

A bold teal-colored infill joins the two sides of the U-shaped building. Materials were selected to complement the existing concrete, limestone and brick of the original facade.

Given the stipulation that the character of the original structure be maintained, materials were selected to complement the existing concrete, limestone and brick, while simultaneously providing the building with a new marketable identity.

On the exterior, the atrium is clad in teal-colored aluminum curtainwall; the window frames are also teal, set against double-glazed windows.

Throughout the building, architectural details create a relationship between the interior and exterior and harken to the building's art deco origins. Painted wood bands along the walls reflect back to those found etched on the stone crest on the exterior, and the teal was carried through to the interior on custom-designed art deco light fixtures and railings.

Details along the circulation axis further reinforce the updated image. While the rectilinear art deco exterior was maintained, colorful curvilinear elements such as the central bay window, the column covers and the curved balconies reinforce the new image of 1010 Metrodome Square.

What originally read as the back of the building became the front entrance after the reconstruction of the former knitting factory.

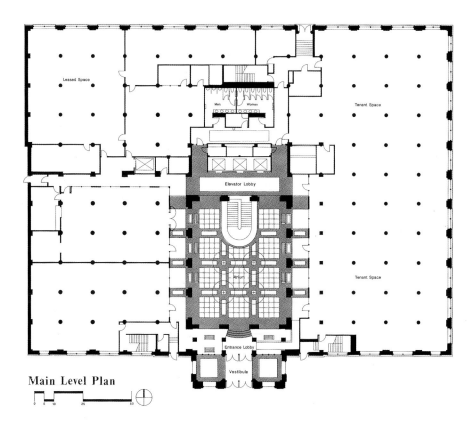

Main Level Plan

The 90-ft.-high atrium spans 50 feet between the two existing wings.

Perimeter detailing, curvilinear elements and balconies help to reinforce the building's new identity.

The architect wanted to avoid designing an uninhabited glass void at the center of the building. So the atrium, capped by an expansive skylight, is filled with benches, trees and plants to encourage public use.

The teal color was carried from the exterior to the interior and appears on the art deco-inspired decorations such as railing and light fixtures.

HTB, INC. OFFICES

location
Oklahoma City, Oklahoma

architect
HTB, Inc.

contractor
Yordi Construction Company

photographer
R. Greg Hursley

Architectural firm HTB did more than merely acquire new quarters when it renovated and relocated its offices in an abandoned church. The building is situated in a once-fashionable neighborhood just outside Oklahoma City's business district and HTB hoped that its move there would be inspiration for further redevelopment in the vicinity.

Two very different renovation goals are embodied in the conversion of the 1907 Maywood Church—creating a technologically modern office, while simultaneously maintaining and incorporating the ruins of the existing structure into the design.

The abandoned church had suffered a number of ill-conceived renovations and the final blow to the much-maligned structure came in the form of a lightning bolt that skewed a portion of the building.

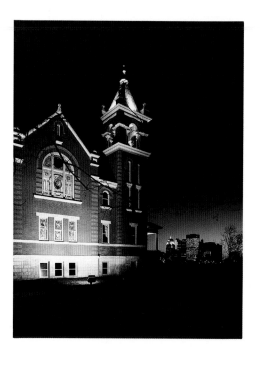

While it would seem like an abandoned 1907 church with serious lightning damage would be destined for razing, HTB architects envisioned the concept of a "building within a building" and incorporated the ruins into its new corporate headquarters.

But rather than razing the ruins, HTB architects envisioned the concept of a "building within a building." Designed in accordance with U.S. Department of Interior criteria that stipulates a clear separation between old and new, HTB designed a L-shaped addition that was wrapped around part of the church.

With gables and a tall arched recess that recall the rounded stained glass windows, the new 29,751 ft. addition echoes the appearance of the old church. While the buff and red brick exteriors of the addition and the church are an exact match, the new building and the remains of the church are clearly distinguished from one another in the interior. Although the two spaces are connected by a skylit atrium, the centrally-situated church ruins stand relatively free and present a unique contrast to the addition.

As if the ruins of a church at its center wasn't enough to lend personality to the new building, the design team also specified some faux painting depicting a Roman ruin and a ceiling mural of the sky. In some spots, the church's original plaster had cracked and peeled, but rather than fixing it, the damage was left intact and serves as a transitional element between the old and new portions. Larry Keller, one of the project's designers calls it, "instant ruin." He said, "We're trying to confront people. We left the old building and that will shake a lot of people up and I'm sure they'll ask, 'Why didn't you finish the old building?' We just felt we could make a much stronger architectural statement leaving it as a ruin."

Between the old and new structures, the church's damaged plaster and brick was left untouched to serve as a transitional element between the two buildings.

A three-story addition was essentially wrapped around part of the remains of the brick structure. The church ruins are the centerpiece of the new building and a glass elevator was ingeniously tucked in its center.

Systems furniture was selected and laid out to follow the horizontal lines of the building.

A view through the old church's arched opening illustrates the unique contrast between the church and modern office. Despite the odd juxtaposition, the concept works both functionally and esthetically.

GRADUATE HEALTH SYSTEM

location
Philadelphia, Pennsylvania

owner
Graduate Health System

architect
Mark B. Thompson Associates

construction manager
Gaudet Associates

photography
Tom Crane © 1989

Like many religious institutions located in urban areas, The Church of the New Jerusalem had lost a considerable segment of its membership when people started migrating to the suburbs. The church found that it was no longer able to maintain its congregation or its circa 1881 structure.

Because the building was situated in a prime section of the Philadelphia real estate market, it was feared that if the structure were sold on the open market, it would be demolished. But, developer Edward S. Brown Group bought the building, acquired a tenant and guaranteed that the character of the English Gothic Revival–style church would be maintained.

Tom Crane © 1989

One would never guess that behind these ornate brownstone walls and stained glass windows is a modern corporate headquarters.

However, Brown Group's tenant, Graduate Health System, required nearly two times the square footage than was available in the church. The significant challenge was to double the floor area without altering the historic charm of the building.

Part of the space problem was solved by constructing a 5,550-sq.-ft. below-grade addition for executive offices and a conference room. Additional space was captured by inserting a new steel frame into the building's shell to support two new floors of office space within the transept. The 75-ft.-high vaulted wood ceiling was retained as an architectural feature in the new workspace. The 65-ft.-high transept as well as the chancel and vestry spaces were preserved and are now the building's atrium.

A custom aluminum-framed glass curtainwall was built to maintain the open feeling within the building and allow the office occupants to overlook the atrium. To provide easy vertical circulation between the office levels and the atrium, a double helix circular stairwell was fabricated in cast aluminum sections and bolted together at the site. Clad in a cherry veneer, it matches the original woodwork of the pulpits and organ casework.

While it was necessary to provide the office with the functional necessities of a modern company, Graduate Health System did not want its decor to clash—or compete —with the Gothic architecture. Soft, neutral colors were purposely selected for carpeting, wall coverings and workstations. The receptionist's desk was fabricated from the cherry wood salvaged from the church's original pews.

Most of the significant architectural features of the original church were masterfully reused when the building was adapted for commercial use.

While it is a rather unusual spot for a corporate headquarters, the former church provides Graduate Health System with an identifiable presence in Philadelphia. The company's managers feel that their space is "more than an anonymous floor in an office tower, and that in an era when modern architecture can alienate, this new headquarters has preserved architecture that is warm, inviting, beautiful and irreplaceable."

Tom Crane © 1989

Tom Crane © 1989

The four stone arches along the east wall and the organ casework and pipes were all restored, and serve as the visual centerpiece of the new reception area.

Tom Crane © 1989

124

This 60-ft.-high glass curtainwall was custom-designed to preserve the open feeling of the church and allow employees to look down into the spectacular atrium.

Subdued interior finishes were selected for the office areas of Graduate Health System to avoid pitting its design against the graceful architectural features of the church.

CHAPTER

III

Retail
Renovation

$\mathcal{M}$ any variables come into play in measuring the success or failure of retail space. After the ever-famous "location, location, location," ambiance ranks among the most important criteria that combine to ensure success in a ruthlessly competitive market.

Retail centers are probably the most trafficked square footage anywhere, and at the same time are expected to be among the brightest, best maintained and attractive, as retailers strive to lure the elusive consumer.

Updating these spaces to keep them in the latest fashion, as well as safe and comfortable for occupants, is the big renovation challenge. Rarely are retail centers shut down, even briefly, to accomplish reconstruction activity. The challenge is intensified even further by the varied tastes of both the consumers and the retailers, and complicated by the budget constraints imposed by owners.

The true measure of renovation creativity is demonstrated by the experts who can successfully accomplish a renovation of retail space that is esthetically pleasing, reasonably within budget, and meets the stringent requirements of both retailers and consumers alike.

The sampling of retail renovations in this chapter, in addition to being diverse, demonstrates some of the latest trends being incorporated by savvy renovation architects determined to give everyone—owners, tenants and retail customers—the ultimate experience in state-of-the-art retailing settings. In every case, the measure of complete success is not limited to esthetics, but incorporates bottom-line effectiveness through increased consumer traffic—and spending!

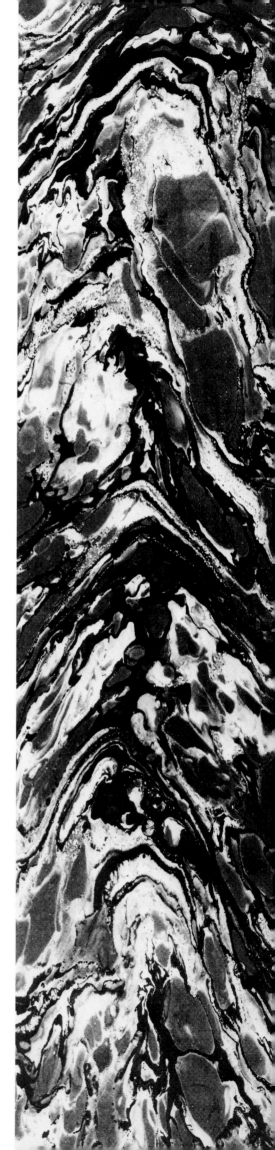

NORTH PIER

location
Chicago, Illinois

owner
Broadacre Development

architect
The Austin Company

photography
**Wayne Cable; Gerald West/The Chicago Tribune
(night shot)**

While the waterfronts of major cities such as Boston and Baltimore were being transformed into entertainment meccas, Chicago's former industrial riverfront, for decades, sat largely overlooked and underutilized.

It now is experiencing a renaissance with the development of Cityfront Center, literally a new neighborhood that will include commercial, hotel, residential and retail buildings. The front-runner of the redevelopment is the seven-story North Pier, a former furniture exhibition and distribution facility, built in 1902, that was converted into a retail and office center.

The development of North Pier transformed a largely ignored part of the city into a shopping, eating and entertainment mecca.

Located north of the central business district at the confluence of the Chicago River, North Pier now houses 280,000 square feet of office and 170,000 square feet of retail space. Chicago developer Broadacre wanted to create a new public gathering place, offering a broad range of activities, while capitalizing on the building's historic character and its location along the water.

With high ceilings, hewn wood support beams and columns, the building was well-suited to a mixed-use development. Its 120-foot depth provided retailers with deep spaces along double-loaded corridors and large spaces at the ends of the building were appropriate for the restaurants that serve as anchor tenants.

A conscious design effort was made to maintain the integrity of the original building and its industrial appeal. Heavy timbers add to the building's visual appeal and interior design elements were selected to complement the warehouse atmosphere. Steel floor tiles, originally used when the structure was a warehouse, were salvaged and incorporated into the floor design. The light fixtures, fabricated of perforated metal, provide uplighting to highlight the original, refurbished ceiling. Signs hung from the ceiling by cables are seemingly towed in by pulleys held by whimsical sailor characters clad in yellow rain gear.

Extensive work was done to the expansive "back porch" at the water's edge to encourage outdoor activity. Though parts of the original building were demolished, beams were salvaged to create two three-story glass galleries overlooking the water. They provide seating for the many restaurants in the building and offer spectacular views of the city's skyline.

Along the street, or "front door," where loading docks previously existed, grand entrances were carved for the retail portion. The remaining loading docks were deftly redesigned into large display windows.

Extensive work was done to the "back porch" to capitalize on the building's location along the water. Docks serve as additional seating areas for the bars and restaurants within the building.

Steel floor tiles, used when the building was a warehouse, were incorporated into the floor design and custom-designed light fixtures highlight the refurbished ceiling.

A three-story atrium at the center of the space unifies the shops and restaurants around a focal circulation core.

Once a furniture distribution warehouse, North Pier now houses 280,000 square feet of office space and 170,000 square feet of retail and restaurant space.

The brick facade, once painted white, was sandblasted to reveal the building's original color. Terra cotta detailing was cleaned and more than 700 new windows were installed to adapt the building for its new use.

CUMBERLAND MALL

location
Atlanta, Georgia

architect
Brian Thornton, RTKL Associates

owner
The Yarmouth Group

photography
Dave Whitcomb

Playing on Southern garden themes, RTKL designed a complete renovation for this 1.9 million-sq.-ft., two-level mall. The mall is anchored by Sears, Macy's, JC Penney, and Rich's.

Completed under budget, the renovation included all new lighting, floor paving and landscaping, as well as new benches, planters and trash urns.

At the entrance, painted steel, lattice-covered pylons with metal banners provide a fresh new entrance approach. The mall was open continuously during the renovation process.

The interior was further enriched by a rust, peach and sky-blue pallette, lush landscaping and a pool that, when covered, doubles as a stage for in-house promotions and other special events and entertainment.

The original open-air elevator was enclosed with glass. Lattice-patterned laser-cut steel accent rails were applied to the elevator, gazebo retail kiosks, and custom torcheres. RTKL also added highly refined copper handrails to the existing glass balustrades.

Painted steel, lattice-covered pylons with metal banners mark the impressive entrance to the renovated Cumberland Mall.

The elevator, which originally was open air, is now enclosed with glass and accented with lattice-patterned laser-cut steel rails.

Focal point of the mall is a strategically placed pool which, when covered, can be used as a stage for in-house promotions, entertainment and other special events.

All new lighting, flooring and interior architectural and landscaping touches have transferred the Atlanta Cumberland Mall into an all-new showcase of Southern "merchandising hospitality."

Gazebo retail kiosks and
custom torcheres are also
accented with lattice-
patterned, laser-cut steel rails,
enhancing the Southern
garden theme of the mall.

location
Atlanta, Georgia

architect
**Cooper Carry & Associates, Inc. Architects;
Turner Associate/Architects and Planners, Inc.**

photographer
Tim Hursley

In the late 1960s, Underground Atlanta was a vibrant and successful entertainment district, attracting local residents and tourists. However, much of its regular clientele was lost in the flight to the suburbs. When the drinking age was lowered in 1972, Underground Atlanta became a collecting point for teenaged drinkers and later, more than one-half of the center was demolished to make way for the city's rapid transit system. By the time it closed in 1982, it had become a low-end commercial district and a cause célèbre of the media as a crime-ridden locale.

For the new Underground, the design team formulated the concept of the "urban village" or "town center," a shopping, dining and entertainment district that would appeal to all segments of the population and draw tourists and convention attendees as well.

"The 'urban village' idea incorporates the amenities of the old central business district, or downtown, into a contemporary structure that creates a community environment. It also preserves the community's heritage and infrastructure," one of the creators of the concept has said.

Preserving the heritage of the city was a primary goal of the project. Wherever possible, the new design incorporated historic remnants such as existing facades and decorative friezes.

In an attempt to weave the center into the fabric of the city—preserving heritage—new construction was merged with existing buildings and other historic remnants such as cast-iron facades, decorative friezes and rubblestone foundations. Two turn-of-the-century buildings were also incorporated into the design of the dual-level, six-block project.

The design team avoided creating definite entrance and exit points for Underground Atlanta to avoid the trappings of the typical segregated urban malls. Allowing access from four directions makes the project a more integral part of the cityscape.

To appeal to a diverse crowd, the merchant mix is varied. Retailers include both local and national stores and the nightclubs offer a variety of musical entertainment. Dining options range from fast food to more elegant sit-down meals. The revived development has three times as much leasable area as its predecessor.

The merchant mix is varied to appeal to a diverse crowd. "Down the Hatch," an eating and entertainment venue had been a popular destination in the original Underground.

To better blend the new Underground Atlanta into the cityscape, the architect avoided creating any definite entrance and exit points.

COLISEUM MALL

location
Hampton, Virginia

architect
Peter Faillot, RTKL Associates

owner
Mall Properties, Inc.

photographer
Dave Whitcomb

In the competitive world of shopping malls, ambiance is often the buzzword. The Coliseum Mall, although not very old, was in need of renovation to give it a fresh, contemporary look.

The regional retail center underwent a complete interior revamp incorporating new colors and extensive use of stylized colonnades, pediments and trellised screens to provide a unifying design theme. The result evokes an updated aura of Virginia Tidewater elegance.

A new food court positions nationally-acclaimed food destinations around a bright skylit garden with trees, flowering plants and vines.

Entrance signage at the Coliseum Mall is distinctive and inviting.

Sparkling pathways to retail shops are created by new agglomerate marble floors in a timeless, geometric pattern. Coupled with dramatic uplighting and expansive skylights, the mall renovation provides a new and invigorating shopping experience.

In addition to the new interior spaces, the mall's overall allure has been enhanced with new entrance signage and interior signage, combined with colorful accent flags and banners.

Dramatic uplighting and expansive skylights give a bright, airy feeling within the mall.

Interior signage and colorful flags and banners provide an exciting, yet versatile unifying panorama to the interior public spaces.

The new food court with its skylit garden is a popular gathering spot for shoppers in the Coliseum Mall.

PAVILION SAKS FIFTH AVENUE

location
Houston, Texas

architect
RTKL Associates, Inc. (Dallas)

owner
Kenneth H. Hughes Interests

contractor
Hayman Company

photography
Scott McDonald, Hedrich-Blessing

Before its renovation, the Pavilion Saks Fifth Avenue suffered from a timeworn image and a lack of patrons.

The renovation of existing space and a 55,000-sq.-ft. addition in front of anchor tenant Saks Fifth Avenue gave the center a stronger urban presence and a fresh new appearance. The addition is comprised of a three-story atrium featuring upscale specialty shops and below-grade parking. A barrel-vaulted skylight now connects Saks to the main atrium.

Using the architecture of Key West (Florida) as an inspiration, the architect selected cut stone, plaster facades and a tile roof to evoke a tropical motif for the center.

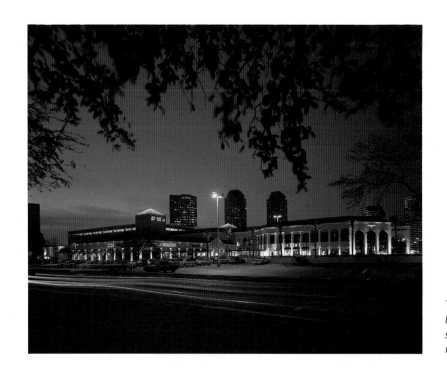

The renovation and addition helped to create a stronger street appeal and shed the retail center's dated image.

New flooring, wallcoverings and finishes revived the cold, unadorned interior. New glass balustrade escalators move patrons between the levels; circulation spines are covered with Arizona sandstone and green marble accents that were salvaged from the original floors. New oil-rubbed mahogany handrails rest on custom-painted ironwork and stone curbing. Special effects lighting and 90-ft.-high palm trees help to reinforce the tropical motif in the interior.

Columns, palm trees and special effects lighting help to invoke a tropical motif for the Pavilion Saks Fifth Avenue.

New finishes freshened the previously cold, unadorned interior space. The new handrails are oil-rubbed mahogany and rest on a base of stone curbing.

A barrel-vaulted skylight
lightens the interior and
further reinforces the retail
center's tropical motif.

Glass balustrade escalators
move patrons between floors
of specialty stores. The
circulation spines are covered
in Arizona sandstone with
green marble accents.

CHAPTER

Institutional
Renovation

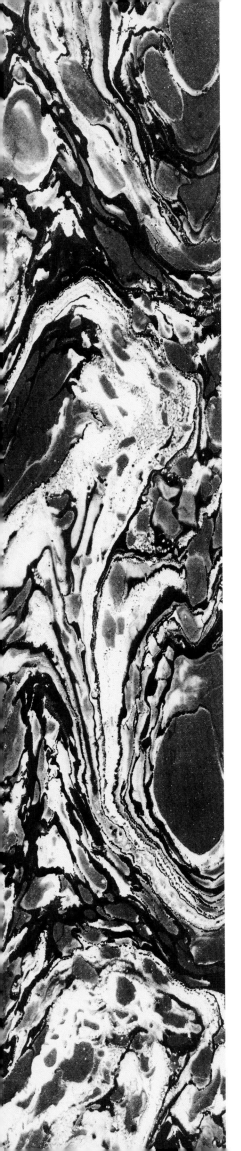

In many cases, the decision to renovate an institutional structure is predestined by the nature of the building. Today, hospitals, schools, churches and other such public buildings are seldom, if ever, torn down and replaced with new structures.

Sometimes, historical preservation directives dictate just how much, and what type of reconstruction can take place. Often, sentimentality comes into play in decisions to renovate institutional structures.

Also, additional definitive descriptions become more prevalent in institutional reconstruction than in any other area of reconstruction activity: descriptions such as "restoration" and "revitalization." Such descriptions are generally used in conjunction with renovation activities on schools, churches and other special-purpose public buildings, but are by no means limited to such structures.

Institutional buildings tend to be much older than other building types which may be renovation candidates. Therefore, the renovation activities performed on these structures are in most cases, highly specialized procedures, procedures requiring gifted craftsmen who are masters at their respective trades.

Furthermore, institutional buildings tend to require specialized expertise to ensure that the reconstruction activity is performed in compliance with unusually stringent building codes and in compliance with predetermined historical preservation guidelines.

This chapter highlights a number of reconstruction projects in the institutional arena which were particularly challenging and which utilized the expertise of particularly talented teams of reconstruction and restoration specialists.

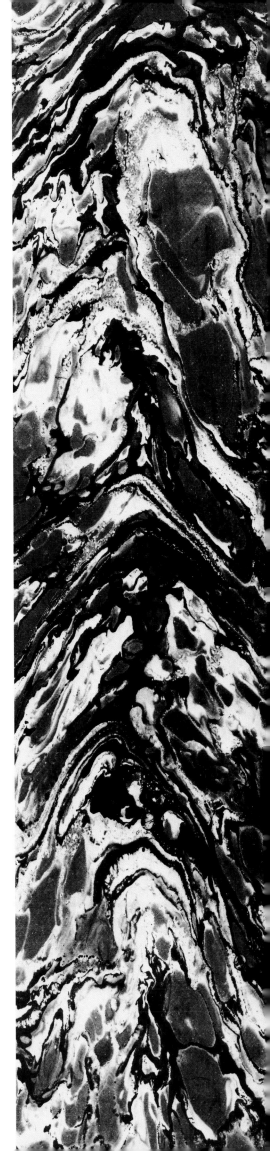

CHICAGO THEATER

location
Chicago, Illinois

architect
Daniel P. Coffey & Associates, Ltd.

contractor
W.E. O'Neil Construction Company

photographers
Don DuBroff (exterior); Barry Rustin (lobby and hallways); Steve Blutter (staircase and auditorium)

As they started the renovation of the Chicago Theater, work crews discovered memorabilia strewn throughout the building that gave them a sense of the structure's history. A note written to Joan Crawford, a 1914 French Foreign Legion pin, a program from 1921 and a newspaper clipping with the headline, "Mussolini Takes Over," were among the items found.

Designed in the 1900s, Chicago's most spectacular theater palace was destined to tumble down had it not been issued a reprieve through the efforts of local historic preservation enthusiasts.

Despite two insensitive renovations that badly damaged the building's opulence, the architect was able to recapture the original splendor that was the building's hallmark. Within 12 months, the cumulative effects of dirt, sloppily-applied paint and false ceilings were erased.

The exterior renovation restored the internationally famous marquee and the "Arc de Triomphe" motif that was executed in cream-colored terra cotta.

The 55-ft.-high lobby, patterned after the Chapel of Louis the XIV at Versailles, features marble, ornamental plaster and scagliola that was meticulously restored. The space is illuminated by grand crystal chandeliers that were rescued from other theaters of the era that had already succumbed to the wrecking ball.

A grand staircase leads patrons to the various levels of the theater and offers them views of the lobby, promenade and ornately-decorated hallways.

In the 3,800-seat auditorium, new paint highlights the rich ornamentation; lost grillwork and decorative pieces were authentically refabricated. Dramatic murals, one of which depicts the Greek god Apollo, were restored using a technique that remoisturizes dry oil paint. The auditorium is illuminated by eight chandeliers and original Steuben sconces that were cleaned and restored. The Wurlitzer organ, one of only five pipe organs in Chicago, was restored by the

Chicago Area Theatre Organ Enthusiasts, Inc.

With new dressing rooms, stage equipment, lighting and sound systems, the Chicago Theater is well-equipped to accommodate future generations of theatergoers.

Before the restoration, workmen found dust, bullet holes in walls, wood panels covering decorative plaster and sixty-four years of cigarette and cigar smoke that had yellowed the marble in the lobby.

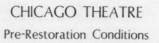

CHICAGO THEATRE
Pre-Restoration Conditions

The grand lobby, patterned after the Chapel of Louis the XIV's palace at Versailles, features marble, scagliola and ornamental plaster.

As patrons scale the grand staircase, they are provided with views of the grand lobby, the promenade and other ornately-decorated areas of the theater.

*The ornamental details in the
hallway had been obscured
behind drywall prior to the
renovation.*

Ceiling murals in the
auditorium were restored
using a technique that
remoisturizes dry oil paint.

THE CATHEDRAL OF THE INCARNATION

location
Nashville, Tennessee

architect
Gobbell Hays Partners, Inc.

photography
Rion Rizzo

The primary aim of the $1.4 million renovation of the 1914 Romanesque-style church was to redesign a number of elements in the building to reflect the liturgical changes suggested by the Vatican II Council, as well as highlight the cathedral's architectural appeal.

A significant program requirement was reconfiguring the space to encourage increased participation by parishioners. To this end, a new central altar replaced three existing ones at the front of the nave and the baptismal font was relocated from a side entrance to the main entrance.

The Cathedral of the Incarnation, built in 1914, serves as the "mother church" of the diocese in Nashville.

Energy efficiency and user comfort were also of considerable concern. The existing heating system was steam-piped by tunnel to floor radiators and the air conditioning system had been installed in the attic. Air was fed through eight supply grilles in the lower side aisle. The new HVAC system is suspended in the attic and sized to prevent noise from entering the worship space. New air supply grilles, cut through the clerestory wall, do not compromise any of the architectural character because they are virtually unnoticeable.

The design solution also enhanced and preserved the architectural appeal of the building. Deteriorated rubber floor tiles were replaced with porcelain ceramic tiles in multicolored, geometrical pattern and the historic Italian marble communion rail was reused to enclose the baptismal font. Pews dating from 1914 were refinished to reveal their oak wood grain and the new altar, bishop's chair and altar furniture were all designed in oak to complement the pews.

The interior was reconfigured to reflect liturgical changes suggested by the Vatican II Council.

*A new lighting scheme
showcases the blue, rose and
gold ceiling and its plaster
relief work.*

Shown here is one of the plaster Stations of the Cross that was restored and repainted.

The Italian marble communion rail was relocated and encloses the baptismal font.

THE NATIONAL MUSEUM OF WOMEN IN THE ARTS

location
Washington, D.C.

architect
Keyes, Condon & Florance

contractor
OMNI Constructions, Inc.

photography
Dan Cunningham; Larry Olsen (stairway detail)

The National Museum of Women in the Arts celebrates the artistic achievements of women worldwide. The new facility, housed in a former Masonic Temple, brings to fruition the dream of the museum's founder, Wilhelmina Holladay. She pointed out that a young boy hears the names of numerous male artists such as Michelangelo, Renoir and Van Gogh, and knows his dream of becoming an artist is possible. But for young girls, she said, ''. . . That's a large part of what's missing. I would hope that we would be able to offer young women role models.'' The institution has the largest single archive of documentation about women artists, including written narratives, videos, photographs and slides.

The museum is located in a seven-story building that was design by well-known Washington architect Waddy B. Wood in 1907. The renovation of the building involved reconfiguring and outfitting the interior to be suitable for museum functions. Clearly the focal point of the building is the marble-sheathed Great Hall (the lobby), with its two grand marble staircases that scale the sides of the walls and lead patrons to a mezzanine level.

By cleaning and repairing the limestone, granite and terra cotta, the building's facade was returned to its original grandiose stature.

To create additional vertical circulation spines, two additional monumental staircases were created at the rear of the building. Openings were cut into three existing concrete floor slabs to install new supports for the stairs and then straight and curved drywall partitions were then built around the openings to create a circular effect. The new Turkish marble staircase connects the mezzanine level with the second and third floors.

The marble throughout the Great Hall is complemented by an extensive use of faux marble on columns and walls. One hundred fifty-nine pieces of molding were attached to the columns and marbleized in salmon and gray to give the effect of a solid two-ft. marble pedestal at the base of each column.

Other work that required a high level of craftsmanship included the application of gold leaf on plaster ornamentation and the fabrication and installation of circular millwork moldings onto the curved walls.

Upper floors of the museum contain exhibition galleries, a library, classrooms and administrative offices. A fifth floor auditorium is one of the building's original features.

The building's exterior is marked by powerful rounded corners, bold pilasters and cornices. The grandeur of the Beaux Arts facade was reclaimed by thoroughly cleaning its limestone, granite and terra cotta.

This former Masonic Temple was renovated to provide more than 68,000 square feet dedicated to displaying art of women worldwide.

Part of the renovation involved some structural modifications to accommodate the addition of an open circular staircase.

A new circular stairway of
Turkish marble provides
vertical circulation between
the mezzanine and the second
and third floors.

Moldings at the base of the columns in the Grand Hall were marbleized in salmon and gray to create the effect of a solid two-ft. marble pedestal.

The most architecturally powerful area of the building is the Grand Hall that is outfitted in magnificent marble finishes. The museum receives additional revenues by renting the room for black tie events.

CAFE BRAUER

location
Chicago, Illinois

architect
Chicago Park District

developers
**The Lincoln Park Zoological Society and
The Levy Organization**

restoration consultant
Wiss, Janney, Elstner Associates, Inc.

photography
Paul Schlismann

Enviably situated in a city park overlooking one of the lagoons of the Lincoln Park Zoo, Cafe Brauer was designed in Prairie School style in 1908 and was opened for many years as a restaurant.

As early as 1969, a Chicago Park District committee was formed to investigate the rehabilitation potential of the building. The committee overwhelmingly recommended that the building be restored and reopened as a restaurant and one committee member prepared a successful nomination of Cafe Brauer to the National Register of Historic Places to ensure the conservation of the building's historic features.

The original 1908 plans of Cafe Brauer proved to be an invaluable resource in helping the restoration team to replicate the original appearance of the building.

Finally in 1988, plans for the redevelopment of the property were devised. The original 1908 drawings that had been stored in the park district's archives helped the architects to replicate the appearance of the original building, including detailing on the masonry walls and the locations and appearances of the skylight and light fixtures.

The masonry exterior walls were cleaned and repointed; the recessed horizontal brick-red joints were carefully reproduced. The manufacturer of the glazed clay roof tiles created a custom glaze and special shapes that helped to recreate the appearance of the original roof.

The most visually spectacular work was executed on the interior. On the ground floor, the kitchen and restaurant were renovated and turned into a cafe and ice cream parlor that overlooks the lagoon and al fresco dining area. Ceramic mosaic floor tiles were added in the cafe and match those that were restored in the first floor lobby.

In the Great Hall on the upper level, the brick walls were cleaned, and terra cotta murals depicting the park's lagoons were restored. Original paint colors on the ceiling and trusswork were analyzed by removing samples and examining them under a microscope to ensure that the restoration would be historically accurate. A new skylight was installed and bathes the Great Hall with sunlight.

The newly-restored cafe, overlooking one of the park's lagoons provides a pleasant dining atmosphere for visitors to the park and the Lincoln Park Zoo.

NEW YORK UNIVERSITY

location
New York, New York

architect
Haines Lundberg Waehler (HLW)

photographer
Jeff Goldberg

HLW was commissioned for a phased, ongoing program to transform a number of older university classroom buildings to more up-to-date teaching and research facilities.

During the two initial phases, the school's chemistry department was consolidated in the campus's main building complex by converting the Waverly Building to laboratory space and renovating existing labs in the Brown Building. A new laboratory suite was also created in the psychology building for a neural science center which will ultimately encompass 40,000 square feet of labs, offices, animal facilities, research and support areas.

So as not to disrupt ongoing academic activities, the renovations were accomplished as simply and as quickly as possible. In some cases, the programming, design and construction phases overlapped. Space was reconfigured, new pathways for utilities were engineered and as renovation progressed, new utilities were threaded from floor-to-floor.

While the cast iron facade of Carter Hall projects a traditional image, the space within the building is a modern up-to-date broadcast studio.

The renovated interiors were designed to harmonize with the buildings that house them as much as possible. Clean, simple labs and classrooms are highlighted by original architectural details that were retained and incorporated into the design.

Within a single summer, the architect also converted space in Carter Hall that had previously been occupied by the linguistics and clinical psychology departments. There, a professional-quality broadcast center for print and video journalism was created where students can practice as broadcast writers, cameramen, reporters and editors. While the modern facilities contrast with the historic character of the building, the structure's original cast iron columns were maintained and serve as key accents in the corridors.

At the new science facilities in the Waverly Building, the timeworn image of the oppressive corridors were erased and replaced with a brighter and more contemporary design.

Flexible modern lab space was created within the envelope of the older structure by reconfiguring the existing space and retrofitting it with new utilities.

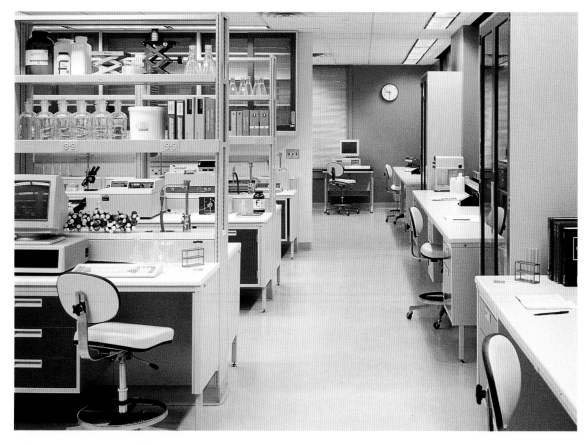

Within the modern space, original cast iron columns serve as decorative accents in the corridors of Carter Hall.

Advanced broadcast facilities provide students with hands-on experience in producing news programs.

CHICAGO HISTORICAL SOCIETY

location
Chicago, Illinois

architect
Holabird & Root

contractor
Pepper Construction Company

photography
**Holabird & Root (before); George Labros
(interiors and daytime exterior); David Clifton
(night exterior); Steinkamp Ballogg (entrance)**

Alternately mistaken by the public as some foreboding municipal building or a mausoleum, the Chicago Historical Society was suffering from something of an identity crisis. While the museum's original 1933 building was a graceful brick Georgian-style design, a 1971 addition, executed in white limestone, was cold and uninviting. One museum official described it as a "neo-Fascist kind of facade."

An addition and renovation served the dual purpose of alleviating a serious storage problem and updating the museum's image.

The three-story addition essentially wrapped around the limestone facade. Its brick sections pay homage to the original building, while a curved glass wall with a white mullion system provide the structure with the desired contemporary appearance. The new construction captured an additional 65,000 square feet for a cafe, a museum shop, administrative offices and underground storage space.

The new addition essentially enveloped the limestone facade. Its contemporary brick and glass design is more sympathetic to the original 1932 museum.

In the lobby, an existing circular staircase was removed and replaced with a grand straight staircase that leads patrons directly to exhibit spaces. Niches along the walls are filled with pieces from the various exhibits.

At the front of the building, elaborate steel trusswork, glass and brightly-colored banners clearly signal to passersby that it is an entrance. Display windows along the facade provide views inside and help to make a lively streetscape and the museum more inviting to the public.

The 1972 addition had an uninviting, "neo-Fascist kind of facade."

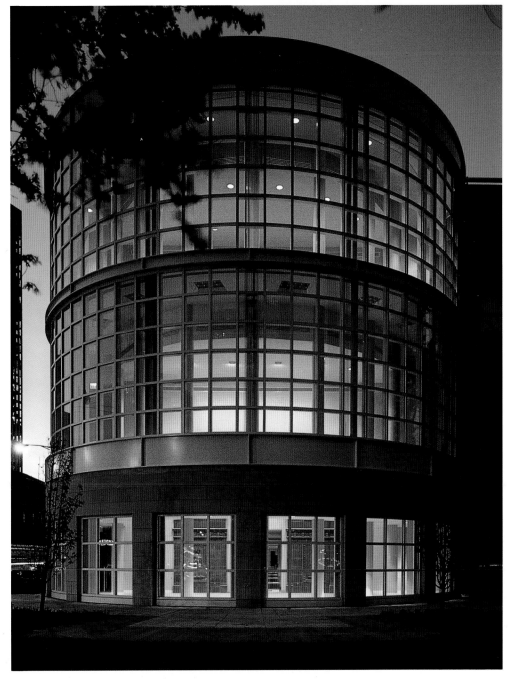

A 68-seat restaurant occupies the mezzanine and first levels behind the curved wall of glass at the southern corner of the museum.

Elaborate steel trusswork and glass create a lively entrance that is clearly distinguishable to passersby.

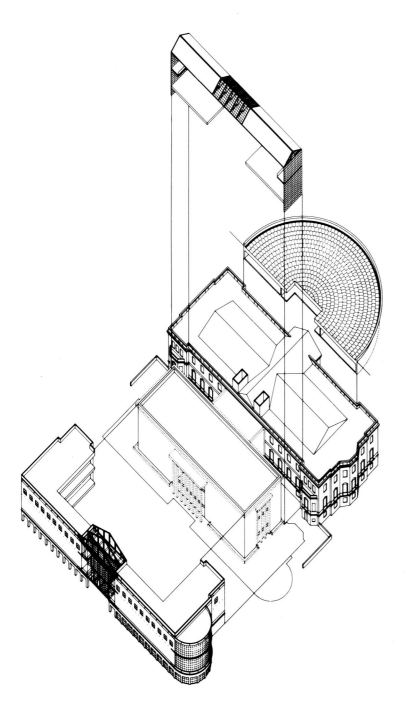

The axiometric illustrates how
the addition wraps around the
existing facade and joins with
the 1932 structure.

A new straight staircase leads visitors directly from the lobby into exhibit space.

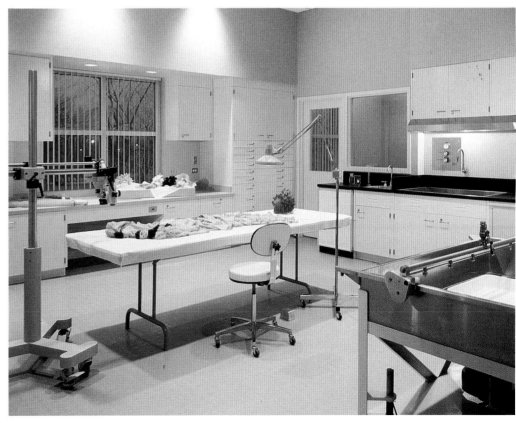

OLD NORTH GYM

location
Lake Forest, Illinois

architect
**O'Donnell Wicklund Pigozzi and Peterson
Architects, Inc.**

owner
Lake Forest College

photographer
**Mark Hertzberg (before); Howard N. Kaplan ©
HNK Architectural Photography, Inc. (after)**

Affectionately known to generations of students as "Old Ironsides," Lake Forest College's North Gym, constructed in 1892, was gutted by a 1969 fire. For subsequent generations, the building merely stood as a constant and gloomy reminder of the unrealized potential inherent in the structure. For nearly 20 years, the building was relegated to a storage facility for the theater department's props.

Given its location in an historic district, it was necessary that the reconstruction maintain the Henry Ives Cobb-designed Romanesque facade of red sandstone. The fortress-like exterior is distinguished by four turrets and a roof with copper ridgecaps and spires.

For nearly 20 years, this handsome building was used as a storage facility for the small liberal arts college.

Movies filmed during the fire assisted the architect in recreating the original lines of the main roof and dormers that had been destroyed. To create additional interior space on the upper floor, the steeply pitched dormers were added back on the new roof which was sheathed in green-blue Vermont slate. New copper flashing, gutters and downspouts were also added.

The exterior belies what one discovers upon entering the building. Originally a one-volume gymnasium, OWP&P created 15,000 square feet of thoroughly modern space on four levels. Two of the turrets were deftly reused as landings for new stairways and nearly all the relatively rectilinear space was usable for offices, classrooms and a lecture auditorium. The third turret adds visual interest to one classroom and a number of offices while the fourth turret serves as a focal point for the first-level auditorium.

New lighting fixtures were selected with an eye toward authenticity. Brass chandeliers hang at the top of the stair landings as well as in the lobby and auditorium. Copper fixtures, reminiscent of those found at the turn of the century, grace the exterior and main public areas. In the stairwells, wall sconces were placed between existing narrow slit windows.

This June 1969 fire gutted the Henry Ives Cobb-designed building.

Two of the four turrets were cleverly reconfigured as stairwells.

Reportedly the formerly one-volume building housed the Midwest's first indoor swimming pool. The new design inserted four levels of modern space within the historic envelope.

Copper light fixtures, reminiscent of a turn-of-the-century motif, were selected to complement the gym's historic flavor.

MILFORD CITY HALL

location
Milford, Connecticut

architect
Fletcher-Thompson, Inc.

general contractor
P. Francini & Company

photography
Robert Miller

While the Milford City Hall is small, only 22,250 square feet, the Georgian colonial structure, with its colonnaded entry and domed clock tower, is nonetheless an imposing structure.

The 1916 building had suffered its share of misfortune, including a sporadic maintenance program, a series of ad hoc attempts at improving utility delivery and a leaky roof and pipes that had gone unrepaired. The final blow to the building's integrity came with a major flood in 1982.

In 1986, the building underwent both a conventional renovation and an historical restoration. On the interior, the restoration brought the central areas back to their original condition. Decorative plaster was restored, woodwork was stripped and refinished and damaged marble surfaces were repaired.

With its colonnaded entrance, tall clock tower and gold cupola, Milford City Hall is a small but imposing building.

Contemporary lighting fixtures were replaced with custom-designed period fixtures and all doors and hardware, some emblazoned with the Milford City seal, were restored and reinstalled. A vast balcony in the city council chambers, closed since 1946 because of structural weakness, was reopened after it was reinforced with structural steel and cast concrete.

Beyond cleaning the brickwork, replacing the roof and restoring the cupola, the exterior renovation returned the building to its original appearance. Using enlarged photographs taken shortly after the building was constructed as a guide, the architect was able to recreate a wood balustrade along the roof line that had been removed and lost, as well as masonry details on the front portico that had been destroyed.

On the more practical side, the building's existing boiler system was converted from steam to hot water and an air conditioning and ventilation system were added. A new transformer was also installed outside the building, and all electrical lines were moved underground.

The renovation has dramatically improved the ability of the government personnel to serve its constituents since they are no longer forced to make do with inadequate quarters and a "patch as patch can" maintenance program.

The Milford City Hall had suffered numerous blows to its integrity, including water damage that destroyed the elegant decorative details on the interior.

Old photographs such as this one, taken in 1916, helped the architect to recreate many architectural details that had been lost over the years.

Based on photographs taken shortly after the building was constructed, the architect was able to recreate the appearance of a balustrade along the roof line that had been lost.

All the windows were replaced with energy-efficient period reproductions. New light fixtures are also authentic to the period.

The restored exterior is dramatically illuminated at night.

Structural reinforcements permitted the balcony to be used for the first time in more than 40 years.

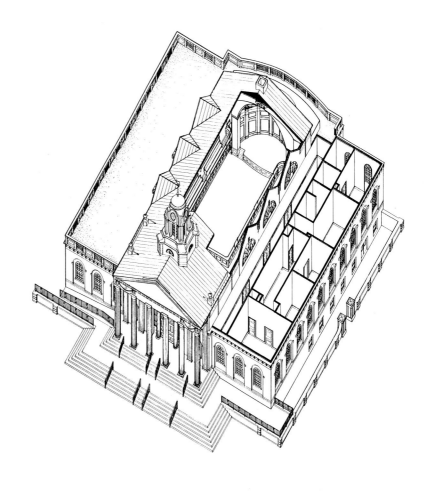

Two grand staircases flank the entrance lobby of the building. The wood doors as well as their original hardware were refurbished and reinstalled.

ST. CLEMENT'S CHURCH

location
Chicago, Illinois

architect
Holabird & Root

photography
Walker Johnson (before); Don DuBroff (after)

For most religious institutions, typically operating on a hand-to-mouth budget, improvements to the physical plant are forced to take a back seat to programs that directly enrich the lives of parishioners.

But the 1905 Romanesque-style St. Clement's Church on the city's north side could no longer postpone its potentially dangerous roof and gutter problems. Water leaking into the church had weakened plaster and caused it to tumble into the pews. Another significant concern was the building's obsolete mechanical and electrical systems.

After the most pressing problems were addressed, a renovation and redesign of the interior was undertaken, which included reconfiguring the space to adhere to Vatican II directives, redesigning the lighting scheme, and restoring the original decorative features.

The addition of copper cladding on the roof was the only significant change made on the exterior of the Romanesque-style St. Clement's Church.

The interior of the church featured a vibrant array of colors and patterns that over the years had been erased and replaced by a bland palette of tan, gray and brown. Elaborate wall paintings, stencil work, painted inscriptions, marbleized surfaces, stained glass and mosaics had all been camouflaged under layers of dirt and paint.

The most significant undertaking of the renovation program was reviving the ornate decorative paint that had covered virtually every inch of the interior.

Based on black and white photos taken in the 1930s, a paint consultant was able to determine where the stencil work had been located. Since color photographs were not available, he was forced to undertake the arduous task of stripping away at the numerous layers of paint to determine the original palette. New stencils were then cut to recreate the patterns and the colors were then reproduced. The photos also indicated that the piers in the nave and apse had featured a rich green and white marbleized finish which was also recreated.

The existing lighting system did little to showcase the walls, the barrel vaults or the dome. To properly highlight the new decorations, the church was retrofitted with custom-designed fixtures that accentuate the elaborate ceiling, piers and altar.

A paint consultant stripped away layers of dirt and paint to uncover the original palette.

The piers that frame the pews were finished in green and white using a marbleizing technique.

Many decorative elements had been painted on canvas and then attached to the plaster. Water damage had caused some of the canvas to become unglued but during the restoration it was repaired and reattached.

CHAPTER

Specialty
Renovation

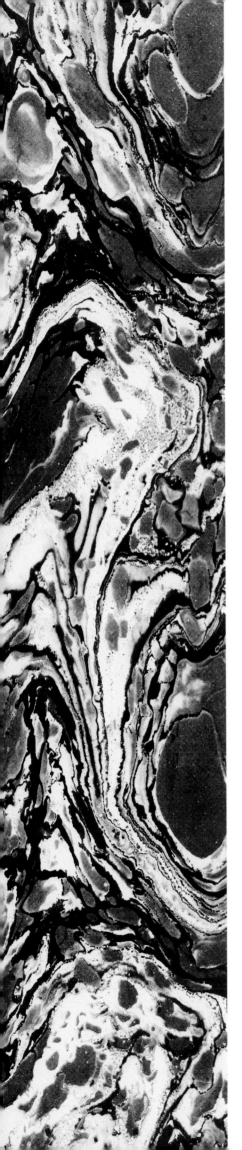

There's nothing quite as exciting or rewarding as what we've come to call "specialty renovation." In truth, the projects presented in this chapter could have been placed within other chapters. However, there were circumstances and challenges that made them each unique within their own genre.

Washington's Union Station can be considered one of the most challenging and rewarding renovation projects to be accomplished in the last decade. An historic landmark which languished in neglect for decades, Union Station was the object of numerous renovation plans. Some never got off the ground; others were aborted. The reconstruction team that finally accomplished the renovation of Union Station is to be congratulated and singled out for special recognition for restoring this landmark to its well-deserved position of preeminence.

The Young Quinlan renovation strikes a high chord for the creativity of the renovation team which transformed this Minneapolis jewel into a viable and attractive mixed-use complex. Restored to its 1920s elegance, yet updated to meet the needs of the '90s, the Young Quinlan project is a premier example of conversion for mixed-use, and a model for renovation teams who have the creativity and patience to pursue this avenue in their efforts to save other vintage structures from the wrecking ball through innovative, adaptive re-use.

Each of the three hotels featured in this chapter has a special renovation story to tell. From the restoration of the huge Intercontinental in Chicago, to the adaptive re-use of the Bellevue in Philadelphia, to the restoration of the Hotel Jerome in Aspen, there are lessons to be learned and accomplishments to be savored.

This final chapter combines with the previous presentations to bring a comprehensive picture into focus, a picture of the magnificent end results achieved by those dedicated to preserving and enhancing some of our highly-prized architectural treasures.

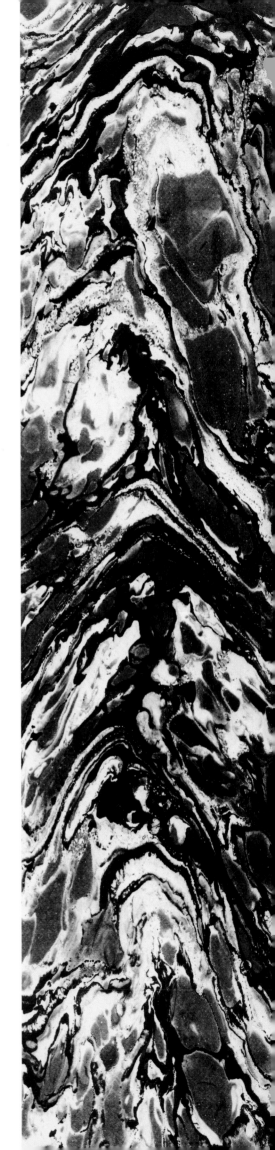

THE BELLEVUE

location
Philadelphia, Pennsylvania

architect
RTKL Associates

owner
Richard I. Rubin & Co.

photographer
Scott McDonald, Hedrich-Blessing

In a grand "rebirth celebration," the Philadelphia Preservation Coalition and the Pennsylvania Preservation Fund held a major gala to unveil the new Bellevue.

The building, originally constructed in two phases between 1904 and 1911 and once the largest convention hotel in Philadelphia (known as the Bellevue Stratford), has been transformed into a major urban mixed-use complex. Following several modernization attempts, the hotel underwent a major restoration in 1979, but closed in 1986.

The challenges addressed in the conversion included preserving all the public and historic spaces, providing enough leasable areas for support uses (such as shops, offices, etc., in addition to the standard needs of the hotel), and eliminating undesirable hotel guest rooms. The new scheme of the mixed-use complex offers a small European-style high quality hotel facility, luxury office space, and elegant shops and boutiques.

Preservation of the exterior of the Bellevue was a major objective in its transformation into a multi-use complex.

The building's upper seven floors house a new 180-room hotel with a spectacular 19th floor lobby. The middle floors provide approximately 275,000 square feet of premium office space. The lower five floors (including one below-grade level) offer 86,500 square feet of retail space wrapped around a central, four-story atrium. An additional atrium, within the hotel portion of the structure, was created through the combination of interior demolition and infill construction within the limits of the building's original light wells.

A major challenge of the renovation was the construction of the 19th floor hotel lobby. A complex system of girders and columns was developed to support the room, known as the Rosegarden, while the atrium was carved directly below this space; the height of the atrium is 85 feet.

In the retail portion of the project, the second atrium was created without disturbing any of the monumental spaces of the structure. All existing "back of house" spaces, such as pantries, kitchens, staff offices, etc., were removed to present the atrium as the focal point of the retail area.

With the exception of the windows, which were replaced following the Secretary of the Interior Standard, the exterior of the building, with its terra cotta fireproofing, was left largely intact.

Entrance to the shops and grand ballroom of the Bellevue signal the renewal of an era of elegance for the former Bellevue Stratford Hotel.

Retention of the natural terra cotta fireproof exterior combines with new signage and other accoutrements to retain the look of prestige at the Bellevue.

The sensitive reconstruction of the Bellevue with an eye toward preservation of details which contributed to making it an historic landmark, are evident in the public areas comprising the lobbies and arcades of the refurbished main floor.

A newly created atrium, one of two, demonstrates the careful attention to details and use of both natural and artificial light to create inviting interior spaces without sacrificing the charm and integrity of the landmark structure.

YOUNG QUINLAN BUILDING

location
Minneapolis, Minnesota

architect
Ellerbe Becket

owner
The 614 Company

photography
**George Heinrich (exteriors, Crate & Barrel,
Polo/Ralph Lauren); Koyama (interiors)**

Uniformed, white-gloved elevator attendants are a fixture of a bygone era. Reinstating the tradition, however, was a natural extension of the renovation that restored the Young Quinlan Building to its smart 1920s appearance.

The five-story building was commissioned by Elizabeth Quinlan, a local fashion buyer who catered to the well-heeled ladies of Minneapolis by providing them with imported ready-to-wear clothing.

Completed in 1926, her store reflected Quinlan's inclination toward subtle elegance: Bronze doors open onto a room with vaulted, ornamental ceilings, crystal chandeliers and 11-ft.-high windows. And at the building's center is a travertine marble staircase with wrought-iron railings.

The exterior is similarly well-appointed: Upper stories are brick and the first floor is covered in Kasota Pink stone.

The building's main entrance features the original terrazzo floor, cherry wood and brass signage.

Given the fact that reusing the store as home to a single retailer was impractical and ill-advised from a business standpoint, The 614 Company determined that a mixed-use arrangement would be the most practical plan for reusing the historic landmark.

Ellerbe Becket redesigned the 155,000-sq.-ft. building to house both retail and office functions. A new central core, incorporating original details from the building, helps to enforce the mixed-use nature of the space, while providing access to the offices on the upper floors.

The open first floor remained as retail space but was divided with corridors that would provide appropriate circulation spines for retailers such as Polo/Ralph Lauren and Crate & Barrel. The second floor was reconfigured for a mix of office and retail tenants while the upper three stories are dedicated strictly to offices.

This original 1926 rendering and a 1980s photograph of the Young Quinlan Building illustrate that the appearance of the structure has not changed since it was first built.

Law firm Arthur Chapman McDonough occupies the fifth floor which was once a warehouse/storage area for the store.

The arched window was personally selected by the founder of the Young Quinlan store and helps to project a genteel, homey image to potential customers. This window now showcases a selection of Polo women's clothing.

The original brass doors at the
Ninth Street entrance were
refurbished and reinstalled
during the renovation.

A more traditional image was desired for the Polo/Ralph Lauren store. The wrought iron railings, chandeliers and plaster moldings that decorate the space were remnants from the original building and have been designated historic.

HOTEL JEROME

location
Aspen, Colorado

architect
Caudill Gustafson Ross & Associates

contractor
Shaw Construction

photography
Pella/Rollscreen;
David Marlow; Charles White

After being veiled under layers of paint for decades, the Victorian opulence of the 48,500-sq.-ft. Hotel Jerome was unmasked, allowing the building to reclaim its position as the "grande dame" of the city. Located in the heart of downtown Aspen, the 100-year-old structure is, as its architect said, "a cornerstone and landmark in the community, not only as a physical presence, but also as a social institution."

The developer realized that a restoration of the hotel was essential for its economic success in the competitive atmosphere of the surrounding resort community. Built in 1889, the hotel is listed in the National Register of Historic Places. Although the three-story structure had been remodeled several times, it had aged poorly and was due for a major overhaul.

An extensive restoration of the 100-year-old hotel reestablished it as the "grande dame" of Aspen, Colorado.

The facade was restored to its original condition through the chemical removal of 40 years of paint and the reconditioning of the wood store front and metal fascia trim. Special attention was paid to replicating the sizes and shapes of the original wood windows to ensure that they would complement the historical appearance of the hotel. The new windows are metal-clad on the outside and pine on the inside.

The restoration also corrected structural deficiencies that had resulted in a floor sag of as much as 10 inches. The elevator shaft was rebuilt to serve the entire building and the cab was enlarged to accommodate handicapped patrons.

Inside, the 27-suite hotel was remodeled extensively and authentically to its original Victorian character. The 1889 design featured an interior atrium that a previous remodeling had closed. In the new design, it was reopened to link the three levels and a new operable skylight floods the lobby with daylight. In addition, the original wood of the front desk, back bars and stairways was revealed from beneath layers of paint and varnish and was refurbished. The original mintone floor tiles, including cobalt tiles with gold, were removed, reconditioned and replaced over the newly leveled floor. New matching tiles were also extended into the lobby.

A chemical wash relieved the exterior of paint that had accumulated over a 40-year period.

Original mintone floor tiles, including cobalt tiles made with gold, were removed, reconditioned and replaced.

Existing ''cell-like'' rooms were replaced by 27 luxury suites and guest rooms.

After being masked under decades-old paint and varnish, inlaid wood mantels, stairways and the front desk were unveiled.

UNION STATION

location
Washington, D.C.

owner
Union Station Redevelopment Corporation

architects
**Harry Weese & Associates, Inc. (preservation);
Benjamin Thompson & Associates, Inc.
(commercial)**

construction management
**Gilbane Building Company; Sherman R. Smoot
Company**

photography
Carol M. Highsmith

To date, Union Station is unquestionably the most ambitious renovation project ever undertaken in the United States.

When Daniel Burnham designed his Beaux Arts-style masterpiece in 1907, the building was a miniature city and at various times housed a bowling alley, a mortuary, a police station, a butcher shop and a doctor's office. In its heyday, it engaged a staff of up to 5,000.

Burnham selected only the most sumptuous materials for the building: all woodwork was solid mahogany, floors were marble and walls were granite. It was a building luxurious enough to make dignitaries such as Eleanor Roosevelt, Mrs. George Vanderbilt, President Truman and King George and Queen Elizabeth, feel at home when they passed through.

But, when air travel became commonplace and the popularity of rail travel declined in the 1950s, so did Union Station. The building began to slowly decay and in an ill-conceived plan in the 1970s, the building was transformed into a tourist information center; it was a bust. Deemed a public hazard, the station was completely closed in 1981. Later, rain damage caused part of the roof to collapse. The building steadily deteriorated into a home for rats, pigeons and toadstools.

The fate of Union Station then became a political football as Congress debated whether to demolish or restore it. The Union Station Redevelopment Act, enacted by Congress in 1981, ensured that the building would be renovated.

To the tune of $160 million, the building was impeccably restored through a unique public/private venture and was reopened in 1988. Union Station is still a transportation center, but it now also features a vibrant commercial mix. The most historically significant sections were modified very little, while other sections were reworked to house new commercial functions, such as a food and retail court and a nine-screen cinema.

Throughout the interior of the Main Hall, the West and East Halls, the Presidential Suite and the vestibules and gallerias, the original surface designs and details were exhaustively restored.

The building's most sizable space, the Main Hall, spans over 140,000 square feet and features a 96-ft.-high barrel vault with a hexagon-coffered plaster ceiling, featuring egg and dart moldings and gold leafing. Situated around the hall's balcony are 36 massive figures of Roman legionnaires.

Given its sheer dimensions and a restoration strategy that called for recreating the splendor of the station down to the most minute details, preserving the Main Hall was one of the most challenging and labor-intensive undertakings of the project. Often there were 400 craftsmen working on the site simultaneously. The most laborious effort involved hand-applying seven pounds of 22-karat gold leaf to each of the 320 ceiling coffers and casting plaster replacements for damaged sections. Removed in 1951 and no longer available in the United States,

the marble for the gray and white floor was imported from Greece and came from the same quarries that provided the marble for the Parthenon. The 24-in.-square tiles, inset with red corner dots, appear as they did when the station first opened.

Preservation specialist Barianos Historical Preservation, Inc. was retained to perform an almost inch-by-inch discovery mission to unveil all the original decorative surfaces. Some of the firm's research involved scouring through black and white photos of the structure that were housed in the Library of Congress archives and examining miniscule fragments of the building under a microscope to discern the true colors.

In the Presidential Suite, a room that had been the president's private waiting room, a decorative fragment of the original surface was discovered behind a light fixture and that tiny piece was used as a guide to recreate the delicate flower design with gold and silver leafing.

The Barianos team also treated the columns in the 3,400-sq.-ft. East Hall with a virtually extinct artistic procedure—scagliola. By pulling threads through a mix of variously-colored plaster, the technique creates the appearance of veined marble.

Other significant aspects of the restoration included cleaning and repairing the 36 plaster cast Roman legionnaire statues and repairing cast-iron grills, train gates, skylights and clerestory windows.

Once a monumental dining room, the East Hall is now one of the building's retail sectors and contains movable display cases.

Statues of Roman legionnaires stand above the Ionic columns and guard the exterior of the Main Hall.

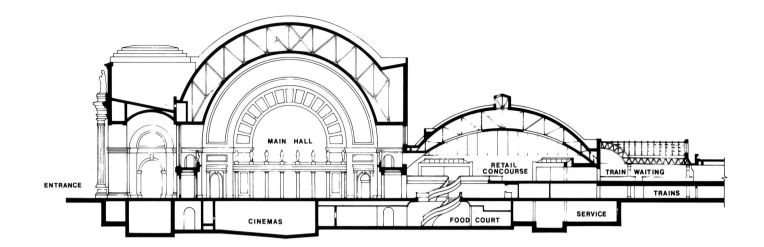

ENTRANCE   MAIN HALL   RETAIL CONCOURSE   TRAIN WAITING   TRAINS   CINEMAS   FOOD COURT   SERVICE

The section indicates where
the new commercial
components were inserted
within the building.

There had been so much water damage in the Main Hall that about 95 percent of the decorative work had deteriorated.

The Main Hall is the building's most cavernous space and required labor-intensive restorative work. Its hexagon-shaped coffers, treated with gold leafing, heightens the magnificence of the room.

Just one example of the lavish, hand-painted stencils that the Barianos team meticulously restored.

The 36 plaster statues that grace the Main Hall were repaired and some received newly-sculpted noses and fingers that had been broken off or damaged. The statues were originally cast as nudes, but during the more modest early 20th century, railroad management feared that the public would be offended and ordered shields that were strategically placed on each statue. The shields remain in place today.

HOTEL INTERCONTINENTAL

location
Chicago, Illinois

architect
Harry Weese & Associates; Design Continuum, Inc.

contractor
Mellon Stuart Company

photography
Wengroff Photography (before); Hedrich-Blessing (after)

In her radio commercials, actress Lauren Bacall describes the renovation of Hotel Intercontinental as "stunning."

But it took a long list of consultants —from historic preservation experts to fine arts conservators—to unravel the mysteries hidden beneath layers of paint, under carpeting, behind walls and disguised under false textures in the former Medinah Athletic Club.

Built in 1929, the building is enviably situated on Michigan Avenue, better known to non-midwesterners as the "Magnificent Mile." The $115 million renovation rediscovered and restored the lavish adornments such as balustraded winding staircases, arched entryways, inlays of marble, intricately detailed brass trimmings and elaborately painted ceilings, murals and frescoes.

Before its renovation and restoration, this 41-story building on Chicago's "Magnificent Mile" was a prominent gathering spot for wealthy locals and travelers during the 1920s.

The original design had been influenced by a myriad of artistic styles, including those from Egypt, Rome, Greece and Spain. Behind the restored limestone facade of the 41-story building lie numerous public rooms all splendidly, and differently, decorated.

The Boulevard Room, considered the most opulent section of the hotel, was designed in the Louis XVI style and was once the formal dining room of the Medinah Athletic Club. The walls are paneled from floor to ceiling with imported Carpathian Elm Burlwood and its walls are topped with an ornate gold leaf cornice that was re-gilded with hand-applied metal leaf.

The elliptically-shaped Grand Ballroom, 100 feet long and 90 feet wide, encompasses two stories and can seat 200 for formal events. Around the perimeter of the ceiling, 37 hand-painted murals on wood panels depict classical landscape scenes. All were restored and remounted. Columns in the room were treated with a faux finish and their capitals were re-gilded.

Above the lobby's grand staircase are the words, ''Es Salamu Aleikum'' that were carved in marble but had been covered under paint. For visitors to the newly-restored hotel, the words are a particularly appropriate message. In translation, they mean ''welcome.''

To bring the building back to its former glory, it first took extensive detective work among craftsmen to uncover the original decorative features that had been disguised under carpeting, paint and false textures.

The Spanish Court is one of many elaborately decorated rooms. The room features an intricately-designed iron light fixture and blue and gold Spanish Majolica tiles.

Artisan Lido Lippi restored 37
hand-painted murals around
the perimeter of the ceiling in
the Grand Ballroom. The
elliptically-shaped two-story
room was described in one
1930 Medinah Athletic Club
brochure as ''a sovereign
setting for balls, parties,
banquets and theatricals.''

Because sandblasting tends to damage many finishes, a unique corn-husk blasting process was used to break up the paint that covered the marble, terra cotta and painted ceilings in the main lobby.

INDEX

PHOTOGRAPHERS

ARCHITECTS

FINISH CONSULTANT FOR PUBLIC AREAS

RESTORATION CONSULTANT

DEVELOPERS

International Hotel Redesign

by Anne M. Schmid

INTERNATIONAL HOTEL REDESIGN beautifully portrays over 40 hospitality projects that have been restored and renovated. Large, lavish full-color photographs provide an in-depth look at each project to allow the designer to see how an area was changed to create a new, updated look or restored to its original elegant state. Before and after photographs, close-up photography of interior and architectural detailing, in-depth captions and project backgrounds provide a wonderful wealth of information.

An insightful foreword by Sarah Tomerlin Lee further enhances this must-read, must-see idea book. Anyone involved in any phase of the hospitality industry or anyone who loves hotels, should not be without this pictorial resource book.

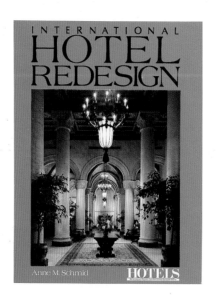

256 pages, 9″ x 12″
Over 375 color illustrations
ISBN 0-86636-113-8

FOR ORDER INFORMATION, PLEASE SEE ORDER FORM ON THE BACK JACKET.

Restaurant Design 2: An International Collection

by Judi Radice

RESTAURANT DESIGN 2 is an international compendium showcasing many exciting and innovative dining environments. Over 300 large, full-color photographs vividly present designs that use state-of-the-art materials and techniques, created by the world's leading restaurant designers.

Many types of restaurants are featured including:

- Full Service
- Hotel Restaurants
- Delis, Fast Food, and Quick Service
- Bars, Lounges and Clubs
- Restorations
- Diners
- Theme Restaurants

...and so much more!

The wide-ranging potpourri of styles and design techniques explored in this valuable, pictorial ideabook makes RESTAURANT DESIGN 2 a "must have" for designers, restaurateurs, graphic artists, and anyone who loves restaurants.

256 pages, 9″ x 12″
Over 300 full-color photographs
ISBN 0-86636-130-8

FOR ORDER INFORMATION, PLEASE SEE ORDER FORM ON THE BACK JACKET.

International Hotel and Resort Design

by Anne M. Schmid and
Mary Scoviak-Lerner

INTERNATIONAL HOTEL AND RESORT DESIGN showcases everything from elegant hotel suites to country-style inns. Over 500 full-color photographs clearly display luxurious rooms, lobbies and conference centers worldwide.

Interior designers, travelers and all involved in the hospitality industry will find this book to be a valuable guide. The interior designs displayed cover varying styles from tropical resorts to Aztec-inspired interiors to the old-world elegance of grand hotels located in Paris, London and Rome.

256 pages, 9″ x 12″
Over 375 full-color illustrations
ISBN 0-86636-068-9

FOR ORDER INFORMATION, PLEASE SEE ORDER FORM ON THE BACK JACKET.